MAY 1 7 2002

Nashville
Public Library
Foundation

*This book
made possible
through generous gifts
to the
Nashville Public Library
Foundation Book Fund*

D1261523

Property of
Nashville Public Library
615 Church St., Nashville, Tn. 37219

Lighting

DISCARDED
From Nashville Public Library

Nonie Niesewand

Lighting

WHITNEY LIBRARY OF DESIGN
AN IMPRINT OF WATSON-GUPTILL PUBLICATIONS/NEW YORK

Lighting
Nonie Niesewand

Copyright © 1999 Octopus Publishing Group Limited
Text copyright © 1999 Nonie Niesewand

First published in the United States in 1999 by
Whitney Library of Design,
an imprint of Watson-Guptill Publications,
a division of BPI Communications, Inc.,
1515 Broadway, New York, NY 10036

Executive art editor **Janis Utton**
Executive editor **Judith More**
Editors **Jonathan Hilton, Stephanie Horner**
Assistant editor **Stephen Guise**
Layout designer **Tony Spalding**
Production controller **Rachel Staveley**
Picture Researcher **Jo Walton**
Indexer **Ann Parry**

Library of Congress Cataloging-in-Publication Data
Niesewand, Nonie.
    Lighting / Nonie Niesewand. — New ed.
        p.   cm.
    Includes index.
    ISBN 0-8230-2775-9
    1. Electric lighting.   2. Lighting, Architectural and decorative
3. Interior decoration.   I. Title.
TK4175.N45   1999
747'.92—dc21   98-41252
                    CIP

All rights reserved. No part of this work may be reproduced,
or utilized in any form or by any means—graphic, electronic,
or mechanical, including photocopying, recording, taping, or
information storage and retrieval systems—without the prior
written permission of the publishers.

This edition first published in Great Britain in 1999 by
Mitchell Beazley, an imprint of
Octopus Publishing Group Limited
2-4 Heron Quays
London E14 9JB

Set in Univers and Rotis
Produced by Toppan Printing Company Limited
Printed in China

1 2 3 4 5 6 7 8 9/07 06 05 04 03 02 01 00 99

Contents

If Stephen Hawking is correct, then two kilograms of matter hurtling about in the void exploded in a flash of light and with a bang big enough to form the planet Earth, along with everything else in the universe. And the rest, as they say, is history.

Light is the powerful medium that has shaped and directed our lives. For the two million years or so of humanity's evolutionary history, our existence has been punctuated by the daily rising and setting of the sun, a diurnal rhythm interrupted only by the feeble glow of firelight during the hours of darkness. Until 1809 that is, when an English chemist called Humphry Davy produced an arc of flame between two carbon rods powered by a gigantic 2000-cell battery. The device was so cumbersome that the world did not switch onto light until Thomas Edison patented his pear-shaped incandescent electric bulb (lamp) and the grid to distribute it in 1879. The first mass-produced light bulb (lamp) was shown at the Paris Fair in 1905 – and the single most important invention in history was launched.

For such a recent invention, electric light has had a dramatic effect on society, revolutionizing the way the world receives information by enabling its delivery through pixels on a computer screen. However, the science of light is still inexact. Whereas music can be written on a score, light cannot be drawn. You can describe sound, but not the qualities and parameters of light. At best it is warm, cold, diffused, kinetic, fragmented, soft, or harsh. Just as there is no geometry to a cloud, so the nearest that light-

optic nerves to our brain, where a colour image of the world is built up. The visual perception of light derives from those electrical impulses travelling from the retina to the brain, where they are downloaded in conjunction with memory and imagination.

But light, natural or artificial, is much more than waves of electromagnetic energy. It carries with it a lot of emotional energy, since light is an indispensable component of both our psychological and physiological equilibrium.

Now that we have colonized the night, creating uniform and evenly lit environments, it is important to talk about darkness. Paradoxically, the new age of enlightenment dawns with more shadow so that we may read our information screens, see the night sky and its galaxies, and experience the sensation of moving between different fields of light. Sir Norman Foster, the British architect responsible for toplighting the cavernous halls of Hong Kong's Chek Lap Kok airport, is more poetic than prosaic when he describes the importance of varying light levels in the way that nature does. "Any engineer can quantify and produce enough light with which to brighten up a passage or by which to read a book. But what about the poetic dimensions of natural light: the changing nature of an overcast sky, the discovery of shade, the lightness of a patch of sunlight?"

Looking at the lighting effects constantly being played out around you helps to place you in contact with nature. The Effects of Light chapter (pages 14–37) considers the effects of natural

Far left The expressive and emotive qualities of light have been charted through the medium of landscape photography. Nature's awesome lighting effects were stilled and tamed in the 1930s and 40s by American photographer Ansel Adams. Geysers spout, tornadoes spiral into the sky, and icy fingers pierce the clouds – all the transitory special effects of light and shade recorded on film in black and white. Understanding the changing patterns of light is as easy as opening your eyes to what is happening around you.

Introduction

ing comes to being scientifically measured is with a scale: a certain quantity of light equals a certain number of lux or candle power. A higher wattage gives a brighter light.

Both Aristotle and Newton were convinced that colour was an integral part of light. We can see these colours only when light is refracted through a prism. This occurs naturally when raindrops suspended in the air separate out the various wavelengths of light shining through them to form a rainbow. Our eyes are sensitive to only some of the wavelengths of light – we cannot see, for example, the infrared or ultraviolet ends of the spectrum. When all the visible colours of light, from red through to violet, enter our eyes, either directly or after being bounced off intervening surfaces, they stimulate specialized cells at the back of the retina. The signals from these cells are transmitted via the

light throughout the day in different parts of the world – from the brilliance of a sunset to the drama of the fireworks of the aurora borealis exploding above you in an eerie silence. Light is infinitely variable, endlessly inspiring. The silvery midnight light in high summer on the Danish peninsula at Skagen can be compared with the Provençal noonday glow, both of which inspired very different 19th-century schools of painting. Artists and architects use both natural and artificial light. The architect Le Corbusier created a spiritual space through colourful light wells beaming shafts of daylight into the cloistered Convent of La Tourette at Eveux-sur-Arbresle, France. A finely delineated crucifix etched in the walls of the Church of Light in Osaka, Japan, by Tadao Ando allows the pattern of the cross to play upon the walls. Angled monumental walls like outdoor screens reflect the

Left Jewellery designer Siegfried de Buck and architect Jan van Lierde call their glass "Grass" lights "garden jewellery". The names of individual lights, which are made by Kreon, are inspired by the botanical names of different plant species. "Festuca", for example, has three yellow-glass shades, the centres of which are powered by two 30–100w incandescent bulbs (lamps). At night, as this photograph by Frederique Debras demonstrates, "Festuca" appears to float on transparent tails.

Right Surfaces change the quality of light, none so constantly as running water. At night, American architect Frank Lloyd Wright's most famous house, "Falling Water", which stands at Bear Run, Pennsylvania, mirrors the light falling on the water. Long and low cantilevered terraces that emerge above the water spread pools of light. At the building's core, fenestrated like a backbone, a column of fragmented light mimics the waterfall. "Falling Water" does not so much take from nature as place the viewer in contact with it.

Mexican sunlight filtered through leaves in a Luis Barragán courtyard. The California-based architect Frank Gehry builds sinuous curves out of titanium to reflect the sky and the water of Bilbao, Spain, to make his Guggenheim Museum a source of national pride.

"For something so powerful, situations for the felt presence of light are fragile" wrote adventurous American artist James Turrell in 1987 in the book *Mapping Spaces* about his underground chambers carved from the Roden Crater in the Arizona desert. He likes to manipulate light so that the spectator feels it physically, to feel its presence inhabiting a space; while artist Dan Flavin used fluorescent strip lights coloured with gels to inhibit the passage of light, stop it dead, and confuse the perimeters by creating a barrier-like wall of light.

Just as there is a distinction between the areas of sky where the sun rises and sets, so there are distinct light cultures in different parts of the world. In the ancient Shoji culture of Japan, a soft, gentle illumination is transmitted through panels of rice paper to illuminate the interior of Japanese homes, whereas in the Middle East light is fragmented through *mous-arabieh* screens. In the Islamic cultures in North Africa and Turkey, light pierces the cores of mosques, its absence as marked as its presence in a play of light and shade that creates powerful patterns. In Byzantine churches in the East, light is fragmented and split by ornately worked screens, while in the West domes allow light to penetrate. In the Pantheon in Rome, the central dome high overhead diffuses the light, giving it a density that is almost tangible.

The basic technologies underlying most of the light sources available today are well established. The first incandescent light was developed in 1879; the first high-intensity discharge light (which suffused cities in its sickly yellow sodium glow) in 1901; the first fluorescent strip in 1938. Lights today are mostly refinements of these innovations, such as the new generation of compact fluorescents that save energy and seek to redress the cool neon light by using tubes with coloured coatings. The miniaturization of the halogen bulb (lamp) in the 1970s produced a new range of architectural installations that made light sources almost invisible. But just as important as the form of the light is the effect the source creates. Even a piece of crystal looks like plastic under neon illumination, but see it under halogen and it sparkles, while under the light from fibre optics it becomes faceted and prismatic.

The physical shape of the bulb, lamp, or strip is determined by the way in which the light source is designed to work. This is why there are so many varieties. Philips alone produce more than 4000 different lights. The light sources themselves can be beautiful objects, though they are often hidden by the fittings. The progression of lighting design throughout the 20th century can be seen in the Hardware Gallery chapter (pages 134–163).

At the beginning of the century, fittings holding multiple light sources harked back in design to oil lamps, candlelight holders, and gas wall-light sconces, but as the 20th century progressed so did light designers. Categories of lights became well defined – wall lights, pendant lights, floor-standing uplighters, downlighters, and spots – and in the last decade the technical change has been so rapid that a new design vocabulary has developed to describe light sources. In Ross Lovegrove's "Solar Bud", sunlight is stored in photovoltaic cells in its torch-shaped head and released automatically at night. Jean Nouvel takes the luminous intensity of special 3M light-conducting plastic film to make an eerie silvery glow of a "Virtuel" light for Luceplan. Ingo Maurer has made it possible to read by a light that has no substance in reality – the incandescent bulb (lamp) suspended above the user's head is a hologram.

How to harness both daylight and electric light at home and at work is the subject of the Applications chapter (pages 62–93). Specialists in the manipulation of light, from 18th-century architect and master of the use of natural light, Sir John Soane, to contemporary lighting designer Philippe Starck, who has a friendly approach to lighting his own home and offices, are illustrated. The difference between the architectural approach, which involves the installation of lights in the fabric of the building, as opposed to the decorator's approach of adding light through

Above Fragmented light is evident in the Institut du Monde Arabe, Paris, by French architect Jean Nouvel. The sunscreen façade has more than 240 glass-and-aluminium panels and 27,000 light-sensitive apertures operating like camera lenses – widening or narrowing in response to changing light conditions.

Left The passing of time is reflected within the interior of a chapel in Tlalpan, Mexico, by Mexican architect Luis Barragán. Traditional materials, such as stone, wooden floors, and stucco walls, are the bold canvas on which Barragán unleashes strips of light, intensified by its passage through coloured glass panes.

Right Light is used as a weapon in the movie *Star Wars*, which reveals the power that light holds over the imagination. In the Middle Ages on All Hallow's Eve, Celts would light bonfires and torches to protect themselves and their livestock from spirits and witches. Now, at night we flood streets with light to scare away muggers.

Right Now that we know light to be electromagnetic radiation, much of its mystery has been lost. Some of the magic endures in candlelight, depicted through the ages in works of art such as *Lesender bei Kerzenlicht* by the Dutch artist Godfried Schalcken.

table or pendant fittings afterwards, is demonstrated by looking at the work of architect, lighting designer, and decorator David Hicks. In one instance he used fibre optics inside his bedhead for bedside reading. Yet in another he controlled light simply by fitting silver and gold semicircles of card inside conventional paper or silk shades to angle the beam. Lighting consultants Sally Storey and Jonathan Speirs with Mark Major use preset, computerized controls to replicate changing levels of daylight. There is much to be learned about using light from trips to the supermarket or visits to exhibitions, galleries, or museums, where problems of up- or downlighting and depth of focus are explored. In supermarkets, light sources change according to the produce they illuminate. At night, the different activity areas of a building are expressed purely through the different layers of light.

Light is as much a state of mind as a technical casebook. Thus, the Case Studies chapter (pages 94–133) explores attitudes to natural light, including the orientation of windows, in interiors by contemporary architects, who make artificial light indoors flattering as well as functional. Light is connected with mood. During the winter months in countries where natural light levels are low, instances of SAD (seasonal affective disorder) syndrome can afflict some individuals, and so emphasis is placed on ways of beaming natural light to the core of a building. In a different context, too much light can be just as bad as too little. Not only is energy wasted when lighting levels are too high, the night sky is often completely obliterated.

The light of the future is intelligent. It can be programmed based on the amount of natural illumination available or it can be regulated depending on the presence or absence of people within different spaces in a building. Innovations in the lighting sector are addressing the design of new light sources, and support technological controls that can be customized. Bill Gates of Microsoft lives in a house that turns on electrically when it is inhabited. Touch screens become works of art or information centres, lights turn on or off as the system recognizes jewellery-like electronic triggers worn by the inhabitants.

Throughout the industry, designers are introducing fittings that can be activated by other than normal switches, including remote control, voice control, programmed systems, and timers. Electronic components are replacing traditional magnetic ones, and longer-lasting light sources reduce energy consumption and help to conserve natural resources. We exist in the world with a heightened awareness of ourselves because the light was, and still can be, turned off, and because we have had, and will have again, the experience of darkness.

Above Fragmenting light gives it a volume. Bauhaus artist Laslo Moholy-Nagy illustrates this quality in the precise geometry of "Lichtrequist". Long before electric light was invented, William Blake's engraving of beams of light symbolized divinity. Moholy-Nagy's graphic seeks to rationalize light as an object, not an expressive force.

Below left The incandescent electric light bulb (lamp) became a symbol for that flash of inspired genius, invention, or bright ideas. Osram's "Nitra" advertisement highlights the basic design of incandescent light, the horseshoe-shaped carbon filament pioneered by Thomas Edison, until it was replaced by tungsten.

Below right Technology has hardly changed from the first light bulb (lamp). A filament, originally bamboo, now tungsten, and a burner made from cardboard cut to about 1in (2.5cm) high, were placed in a vacuum-sealed glass envelope. As a current was passed through, it heated to a brilliant whiteness.

Right A world first is the hologram pendant light that you can read by. A virtual light bulb (lamp) hangs in an empty space. Above it, a halogen source is hidden in a profiled fitting. Ingo Maurer calls the light "Edison, where are you now that we need you?", in homage to the inventor who gave us so much that life would seem impossible without him.

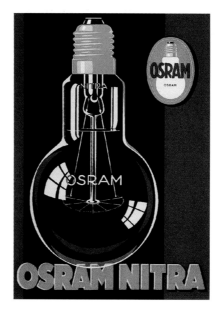

THE EFFECTS OF LIGHT

Tradition meets technology

The world's first "Wire-less" light (above) is recharged in the same way as your mobile telephone and then lights up for six hours. It may look like an age-old Chinese lantern, but the technology that powers it is borrowed from the electronics industry. Another version comes packaged in a wok to symbolize the freedom of movement that these lights guarantee. Andrea Branzi's lights (left) use modern technology to rediscover values that have been lost in modernity in order to cast a warmer, more comforting, and softer glow. The light source is shaded in rice paper called dacron, which is as tough as a cotton jacket, in homage to the softer illumination of the Far East where light is diffused through handmade paper called *wasabi*. B & B Italia launched Andrea Branzi's light fittings on their furniture stand at the Milan Salone del Mobile in 1997.

The most influential architect to work in this century, and one of the first to be switched on to electric light, Swiss-born Le Corbusier, described a house both as "a receptacle for light and sun" and as "a machine for living". He recognized that the expressive value of light could make modern architecture an emotive experience. Light and its special effects were seen then, as they are now, as being both physiological and psychological in nature.

Fairy lights strung up through a tree are festive, but floodlights beamed from a security fence are not. Light can cure and heal, yet it can also be an instrument for torture. Light – and its absence – depends on how it is perceived by the individual. Light is never formulaic. You can measure the distance and beam width and colour temperature of light, and draw up charts of its performance ratios. But the magical ability that light possesses to shape and define space cannot be quantified, since it is entirely personal. That is its power.

The century that began by illuminating the world with artificial light ends with the realization that we need more shade to allow us to see our computer and television screens. Darkness gives rhythm to our existence. Since pagan times, summer and winter solstices have been celebrated. When the midsummer sun never dips below the horizon in Scandinavia, bonfires glow all along the coast. At Diwali in India, clay lamps burn outside every household to celebrate the harvest and the life-giving sun.

Such is the power of light that we even arm ourselves with it. Just as our ancestors back in the Middle Ages lit bonfires, torches, and jack-o'-lanterns to ward off malevolent spirits, so we today floodlight our city

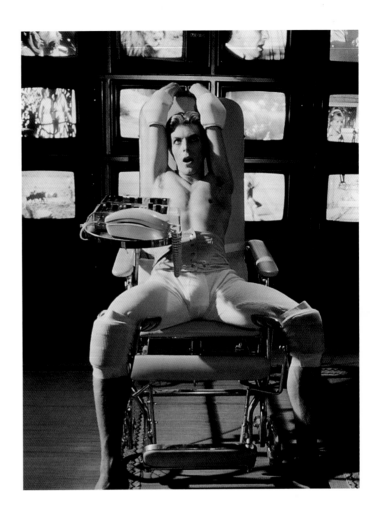

Light and human vision

streets at night to combat crime. In George Lucas's *Star Wars* trilogy of the 1970s and 1980s, the evil Darth Vader was eventually vanquished by the use of a light sabre.

By looking at how light appears in nature, films, paintings, and buildings we discover how different light affects us and how we can manipulate it. "The sun never knew how wonderful it was until it fell upon the wall of a building" American architect Louis Kahn said earlier this century. In different parts of the world the effect of light is expressed differently in buildings and in paintings. The assertive, penetrative light of the Middle East is pierced by lattice screens and falls in ever-changing fragmented patterns that chart the sun's path. The calming glow of light diffused through paper has existed for more than 800 years in Japan through the Shoji culture. In the West, the predominant overhead lighting – through domes and now atriums – balances light with shade.

Left Light fittings in the "Metamorfosi" system bring nature's special effects indoors. "Human Light" is what Italian lighting company Artemide calls its research project into daylight and the seasons in a quest to create an ever-changing poly-chromatic light for interiors. It is a visionary lighting system that will put designer-packaged lights into museums alongside typewriters.

Above David Bowie is *The Man Who Fell to Earth*, but could he read his monitors in a bright interior? Now that information comes in pixels of light to your computer and television screens, what is required is not so much light but shadow. Suburbs have a blue glow at night from all the television sets. Paradoxically, the New Age of Enlightenment dawns on a dimmer world.

In the late 20th century a fundamental shift occurred in the role of light as the symbol of enlightenment. Ever since the 17th century light has been seen as the triumph of reason over ignorance. In the West we still say that we have "seen the light". Now that we need shadows to read the information that is transmitted in pixels of light onto our computers and televisions, shade is more important than bright light. Modern architects, born in an age that believed in a strong diffusion of light, have had to dim the lighting levels so that we can read our screens.

More than a philosophical issue, light levels measured in lux have had to change. In the United States in the 1980s, 1500 lux was specified for the working environment. By the mid-1990s that had dropped to 800 lux. In the UK, 750 lux in the 1980s dropped to 500 lux. Apart from an overall lower level of light, areas of brightness and shade have had to be contrasted to enable us to focus. Discontinuity of light, not continuity, is now important.

Physicists have the answers. Since light needs shade in order to be perceived, bulb (lamp) manufacturers use laboratories to test levels and to balance light with shade. For example, the German manufacturer Erco has chambers, known as dark-cell photometers, to test light intensity where bulbs (lamps) are tested over 24 hours with sensors at certain temperatures to measure heat. A hydraulic ceiling moves up and down horizontally to allow Erco technicians to check beam and distance. Wide beams – like the sun's rays – create background light, spots signal passageways, and floodlights emit homogeneous light from ceiling to floor. Hard, shiny surfaces – white walls and metal stairs, for example – can be warmed up in this play of shade and light. Nature also has a role.

Light and shade

Above Light and shade can highlight cultural differences. In Arab mosques daylight is filtered through latticed *mous-arabieh* screens, or via myriad small openings that pierce the thick clay walls of the Al Janad mosque in Yemen. In Japan, daylight is filtered through paper shoji screens to cast a diffused glow. Light and shade are contrasted more powerfully in the Middle East than in the Far East.

Right Shade accents the Gothic architecture in a cathedral in the north-central region of Île-de-France. Natural light beamed down through stained-glass windows is diluted when it reaches the core of the cathedral to be framed in the ribs of the flying buttresses. Light delineates the huge rib vaults of the skeletal structure to create an emotive space that accentuates its sense of spirituality.

There are rooms that balance natural light from glass ceilings with artificial light. Halls of mirrors reflect horizons – like the sky – with clouds, or a dark night. Parabolic lights balanced with halogen spots connected to computers simulate the position of the sun at any time of the day and season. Think of New York City and the impact the shadows of the skyscrapers have on the streets below as well as on the interiors of tower blocks. At Erco, they can measure that impact.

Light and shade also have an expressive and emotive charge that is harder to quantify. The contrasts between light and shade expressed in Gothic or Roman cathedrals transform them into spiritual places. Reviewing a collection of Lucien Hervé's photographs of Cistercian abbeys, the architect Le Corbusier wrote: "Light and shade are the loudspeakers of this architecture of truth, tranquillity and strength. Nothing further could add to it."

Left The architect of
the German Pavilion
at the Barcelona
International Exhibition
in 1929, Mies van der
Rohe, quoted
St Augustine's "beauty
is the radiance of
truth" as a motto
for architecture. His
glass pavilion is still
regarded as one
of the masterpieces
of modernism. But
modern architects,
born in an age that
believed in such
strong diffusion of
light, as in this
example, would have
to lower the light
levels considerably
today so that
television and
computer screens
could be read within.

Above Architect John
Pawson seeks the
utmost simplicity in his
minimalist designs.
Light fittings set into
the treads of this
staircase at the
Jigsaw fashion store
in London bathe the
white-plastered walls
in light, illuminate the
stairs, and cast an
interesting ziggurat
of shade that leads
upward. When there
is nothing in a room,
Pawson believes that
the exact position of
the light switch
requires the most
careful consideration,
otherwise it might
disrupt the proportions
of the wall.

Cool and warm

Light has an expressive value that can be clearly observed in the difference between cool and warm light. The contrast not only shapes space, it also has the power to brand it. The New-York-based architect Peter Marino identified two very different looks in Donna Karan's two London, England, stores, near one another on Old and New Bond Streets, that are created by using warm light and cool light. The one shimmers, the other glows. The DKNY ready-to-wear label, bathed in ice-white bright light, is the "new kid on the block. I threw so much fluorescent at it that the lighting designer nearly had a coronary" Marino confesses. In contrast, Donna Karan's couture collection is intensely lit with a golden light. In both, the spirit and the energy of New York City are beamed into the gray, northerly light of London.

Cool and warm lights are geographical distinctions in nature's colouring book. Further south on the globe, the light warms up with the angle of the sun. Sicily, for example, has a very different light than Scotland. The great British landscape painters distinguish themselves from their Continental contemporaries in that the weather is not the atmosphere of their pictures – it is the subject of their paintings. The French painter Nicolas Poussin painted landscapes in a clear, even climate; English artist JMW Turner in taunting, flashing, glorious light.

Although there are at least 15 colour temperatures of fluorescent light, this source tends to be cool, while incandescent is warm. Lighting in public places is often a mix of both. Supermarkets are blandly lit with energy-efficient fluorescent light. But fluorescent registers high on blue tones and tends to render reds purple, which is why incandescent light is used over the meat counter. Halogen, launched in the 1970s, has revolutionized lighting. Halogen casts a cooler light than the incandescent bulb (lamp), which is its prototype, but sparkles in its quartz casing.

Left Donna Karan's two London stores by Peter Marino Architects are lit as differently as the sun and moon. Ready-to-wear DKNY (far left) shimmers behind a transparent-glass street elevation. At the rear, backlit walls illuminated by fluorescents behind opaque glass bathe the interior in white light. Banks of AR11 Hollywood floodlights overhead wash the four storeys in a cool light. Video displays add a blue tinge. In contrast, the couture collection of Donna Karan's fashion label (left) has what the architect calls "a celestial light quality". Velvet, silk, gold, black, and intense colour are highlighted by low-voltage projectors screened with GOBO filters that dapple light and shade like sunlight through trees.

Right Light shapes the space indoors and outside Sir Norman Foster's Hong Kong and Shanghai Bank. On the street elevation outdoor light communicates different parts of the building. Metal joists are illuminated and harbour views blacked out. Inside, too, the light is layered. The 177-ft (54-m) banking hall is lit from the ceiling with a large Erco fitting for 570 bulbs (lamps). Seen reflected on a prismatic glass surface, they cast the same light as a 100-lux source on the faces of people below. At table-top level, low-voltage halogens are used.

Northern and southern light

Skagen, on the northerly tip of Denmark, was a favourite resort with Scandinavian painters at the start of the 20th century. So was Provence in France with the Post-Impressionist painters. The chromatic effect in Vincent Van Gogh's *The Night Café* (above) and the cool, luminous light of a Scandinavian *Summer Night on the Beach* (below), by Danish painter Peter-Severin Kröyer demonstrate the power of chiaroscuro in a warm palette and a cool one. Van Gogh's green baize billiard table, warmed by the glow of a pendant light, casts its shadow on a pine floor. The red walls glow. Even a white shade on the light appears yellow. Bathed in the silvery light of midsummer, when the sun never dips below the horizon, Kröyer's picture vividly contrasts the difference in the far north. Just as the pink glow suffusing the ribs in the sand is not enough to take the edge off this cool northern light, so, too, would warm coating and filters on fluorescent light fail to transform it into anything other than cool.

Left The 20th-century architect Le Corbusier, believed that "Architecture is the masterly, correct and magnificent play of masses brought together in light." His chapel of Notre Dame du Haut, Ronchamp, France, carries an emotional punch for everyone who visits it, which is all the more astonishing from an agnostic. Natural light filtered sparingly into the space in light tunnels is poly-chromatic – the modern equivalent of a stained-glass window. As the sun moves, the coloured light falls in patterned slabs on the limestone floor. Light, at once real and immaterial, becomes the symbolic substance of the spiritual. Early modernism offered cold light and hard surfaces – surfaces that Le Corbusier coolly ignored to bring spirituality back into rationalism.

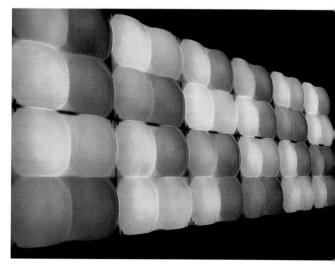

Left Light roller-coasters over the fins of Australia's Sydney Opera House, by Danish Bjorn Utzon with civil engineer Ove Arup, to illuminate its volumes. The building has become an icon – perfect and harmonious as a shell, the exterior is a result of the interior vaulted spaces. Accenting those shell-like forms in bands of coloured light accentuates its organic profile against the velvet dark.

Right A pulsating colour chameleon wall, "Chromawall", by Jeremy Lord is a modular system made of aluminum blocks with polycarbonate fronts that radiate light from four 1¾-in (4.5-cm) golf-ball bulbs (lamps). They change colour with controls that dim and brighten the output. Nine modules bracketed together measure 4¼ft (1.3m) square, with a depth to the bowed front of 9in (23cm).

Colour transformations

An old electricity power substation (left), as large as a city block, in the heart of Edinburgh, Scotland, has a colourful transformation by lighting designer Jonathan Speirs. What was once a threatening industrial plant – painted battleship gray and fenced with skull-and-crossbones "Keep Out" signs – is now a colourful kinetic sculpture. Coloured in the "thistle" greens and purples of Scottish Power, a computerized light program bathes it in red and pink on St Valentine's Day and blue and white on St Andrew's Day in honour of Scotland's patron saint. Seven gigantic ceramic transformers (below) light up the site at night. "There is a poetic sweetness to illuminate a piece of functional industrial equipment that transforms high-voltage electricity," says Speirs, who is now lighting up vast mineshafts in the Ruhr, Germany.

Coloured

Early Christian art depicted light as being red. The light of Creation itself as represented in a mosaic at St Mark's basilica in Venice, Italy, is also red, separated from darkness, which is shown as dark blue. We know from seeing light from the sun pass through a prism, or when it is refracted by water droplets in the air to form a rainbow, that "white" light consists of a blend of wavelengths, each one of which is perceived as being a different colour when separated. Of all the electromagnetic energy radiating from the sun, the visible spectrum is only a narrow band – on either side are ultraviolet, infrared, radio waves, X-rays, and so on.

In order to colour light, you can use filters or a shade. If you admire the colours of a sunset, for example, bear in mind that its chromatic impact comes from sunlight shining through a large volume of atmosphere and being scattered by particles suspended in it. Toward dusk, most of the short blue wavelengths are removed, leaving predominantly the longer orange and red ones. Coloured silk or paper shades on light sources when used in the way decorators dot them about a room cannot hope to replicate this type of special effect.

One way to quantify colour is to use the Kelvin degrees colour index, but its function is to differentiate warm and cool light, not to measure temperature. To understand this, imagine a black object being heated – at first it glows orange and then red, while at higher temperatures it changes to blue and then white. So, cool-coloured light has a high Kelvin rating while warm-coloured light has a low one. Daylight at dawn and dusk has more red and less blue, so its light, like that of incandescent bulbs (lamps), has a Kelvin rating of about 1700. At noon, when blue predominates, the Kelvin rating is higher. Fluorescent light, strong on blues, thus has a higher Kelvin rating of 3000. Halogen scores the same, but its quartz casing gives it sparkle and refracts the light so that we see the colour spectrum around it.

Left Italian fashion designer Romeo Gigli made his first lights for Ycami that cast a rhythmic pattern of light and shade – in the same way that latticed *mous-arabieh* Arab screens do – by piercing aluminium shades.

Right New York City's most heroic period, from the 1920s to the 1940s, depicted in films such as *Dick Tracy*, has street vistas like this work by English artist Christopher Nevinson. Set-back skyscrapers tower above street after street of early-20th-century brownstones, each punctuating the skyline with little grids of light. Reduced to vibrant primary and a few secondary colours under the sulphuric glow of low-pressure sodium street lighting or the blue tinge of the mercury vapour lights around at the time, the reds, yellows, and blues all have the same intensity.

Right The low-budget horror film *Shallow Grave* was a smash hit in Britain in the mid-1990s. Kave Quinn, the film's set designer, used light and colour to turn the film nasty. Not only does the plot thicken, the colours do, too, as they shift from bright twilight to night light. Shadows lengthen. And one of the occupants of the apartment lurks in the loft and drills holes through the ceilings in order to observe the goings-on in the rooms beneath.

Left Portholes inset with halogen spots pierce the paneling of Belgo's basement restaurant in London's Camden Town, making the gray walls three dimensional, like a ship's bow. Architect, furniture designer, and professor at the Royal College of Art, London, Ron Arad made the volumes overhead even greater with trusses stretching out from the cladding across the roof. Concealed behind the cladding, a line of fluorescents bathe the wall below in a continuous, even light.

Right Finnish architect Alvar Aalto was a lighting genius, manipulating light from many sources and angling it to beam it into the core of his buildings. The play of light makes a pattern on this plain, early modernist interior.

Pierced

Divine light is perceived in Western and Middle Eastern countries as entering like a water course, with a pure, undiluted energy and a precise geometry. The English poet and artist William Blake's illustrations of Dante's *Divine Comedy* show the River of Light as water flowing from God to signify enlightenment. Inside mosques, natural light enters through very small openings that pierce the walls so that there is a reverberation within. The light is discontinuous. Narrow beams of natural light that pierce the inner core in controlled amounts cast patterns of shade and colour that follow the path of the sun.

When light breaks through the chinks of the screen from the outside, the rays of light themselves actually become visible. You perceive it as volume. Arabic screens allow light to have a strong presence. Pierced light is neither modified nor modulated. The light it casts becomes emotive and expressive, very different from the evenly lit calm of the light culture in Japan, where, since Shoji times more than 800 years ago, paper screens have diffused light. In the West, light penetrates buildings from overhead domes or skylights that stream and shape the light. If you had to design the density of light, you would observe that, from the early 18th century, light in the West was defined by windows, which let in a watery illumination. In the East, the density was moderated by paper screens and shades.

Light controlled in this way is more violent than other forms. Pierced surfaces, though penetrable as a barrier, can still act as a defence when manipulated by skilful practitioners. In the film *Shallow Grave* the setting changes from an affluent Edinburgh apartment, styled like a Habitat catalogue by designer Kave Quinn, into a terrifying fortress when the paranoid protagonist drills holes through the ceiling to pierce the rooms below. Gotham City, where Batman hung out, is depicted as a gloomy, nightmarish place with skyscrapers spilling discontinuous light onto the streets.

Light that defines

Dynamic Hong Kong by night, seen (left) from the top of the Peak and (right) at street level from Nathan Road. One of the most dazzling skylines in the world shows fragmented light reducing the mass and volume of masonry, glass, and steel that rings the harbour in a kinetic band that snaps on and off. Light splinters the sky and illuminates the night with Chinese characters in coloured neon and presses along the waterfront in serried spikes and jagged curves. To chart a light path from the Peak, pick out the diamonds of light on the Bank of China building by American architect IM Pei on the right. Pei designed this building like sprouting bamboo with nodes signifying growth. It unfolds in storeys as gracefully as a piece of origami. Next to it is the Hong Kong and Shanghai Bank by English architect Sir Norman Foster. Built in gray-clad steel cast with great cross-girders, it is lit by the German-based company Erco to emphasize weightlessness and fragmentation.

Fragmented

The sensation of gazing up at a darkened sky that pulses with countless stars is fragmented light at its most awesome. Where light creates interesting streetscapes in illuminated cities built on a grid system, pattern-making like an artificial starry sky results. Fragmented light, like pierced light, is discontinuous. It differs from pierced light because it is less rigid, more fluid, and on a larger scale.

Early Christian art made light visible through the use of mosaics. Byzantine mosaics, with their metallic tesserae, are examples of fragmented light frozen in stone. The 6th-century poet Paul the Silentiary describing Sta Sophia in Constantinople (modern Istanbul) remarked that "curious designs glitter everywhere". In the 2nd century, the Greco-Egyptian astronomer and geographer Ptolemy identified a cause of optical fusion in colour perception: the angle of vision formed by rays of light from very small different patches of colour, too small for them to be separately identified, seem together to be the same colour. To read fragmented light in cities you need to be guided along specific pathways, like reading the constellations. If your eye tries to take in the whole scene, the flicker and dazzle effect makes it difficult to see.

Once a building is lit up at night it often takes on a different spatial presence than it had during the day. Many glass-fronted buildings are invisible blocks by day but emerge as huge objects of light at night, strong and powerful. We live in a lightscape that is impossible to turn completely off – thus, light is also a pollutant. Legislation designed to reduce light-emission levels began in the 1980s with a pressure group, the International Dark-Sky Association, which seeks to ban the most invasive public light emissions. Once more, the Milky Way hovers above the 82 American cities that have imposed these bans. Ikea advertisements in the 1990s point out that "lower-watt light bulbs are more flattering".

Left An aerial view of Los Angeles in 2019, from the sci-fi film *Blade Runner*, directed by Ridley Scott, as an air car approaches the Mayan-inspired, pyramid-shaped headquarters of the Tyrell Corporation – makers of artificial humans called "replicants". LA in 2019 is a crime-ridden city overpowered by skyscrapers and polluted with light. In the streets below neon reflections shine in puddles of acid rainwater.

Right Bathed in electric-blue light, London's best-known building, Lloyd's of London Insurance by Richard Rogers Partnership, is never perceived as a whole at any one time. Rather, it appears segmented, as if it had been assembled from a kit of parts.

Left Light generated by dots or pixels on the screen makes computerized templates that are ideal for the composition of mosaics. Architect Alessandro Mendini used computer-aided design technology for his digital composition at the Milan showroom of Bisazza, the Italian tile and glass tesserae manufacturer. "The pixellation of the scene creates the same pointillist effect as mosaic," he says.

Right *Ad Parnassum* by the Swiss artist Paul Klee fragments colour and light into squares in a composition that is a vehicle to carry thoughts and forces.

Above The Reichstag building in Berlin, completely under wraps and bound with ropes, was one of the most arresting spectacles of the mid-1990s. In a feat that required great technical and logistical skills, experimental artists Christo and Jeanne-Claude encouraged a worldwide audience to look differently at this monumental slice of history. The eerie blue light that bathes the building's roofline seeks to flatten the folds, seen more distinctly by daylight, into a ghostly shadow of its former substantial self.

Right For about 800 years, sliding paper panels made from a fibrous plant called *kouzo* have diffused daylight entering traditional Japanese houses. Today, these symbols of Shoji culture have been adopted by designers working all over the world. This London house, created for film producer Walter Donohue by architect Jonathan Gale, allows light to pass while safeguarding the occupants' privacy. Gale studied in Tokyo for two years after graduating from university.

"**D**emand concealed or diffused lighting" 20th-century architect Le Corbusier advises in a checklist for the perfect dwelling, in his book *Towards a New Architecture*. Another "must-have" on Le Corbusier's checklist is a sun-facing bathroom, in which one wall is to be entirely glazed, and opening, if possible, onto a balcony for sunbaths. The architect wrote this just 18 years after the American inventor Thomas Edison had presented the incandescent bulb (lamp) at the New York Fair in 1905, and already the need to tame the pure energy and diffuse it, rather than shade it, was firmly in the architect's mind – though obviously not the need for privacy when bathing.

Diffusing tends to scatter and mix the rays of light, spreading them over a wide area, filtered, wrapped, veiled, half in the shadow for a softer, more livable effect. A cult book for the 21st century, *In Praise of Shadows*, written by Junichiro Tanizaki in 1933, explains the different light culture to be found in Japan, where light is diffused through Japanese-made paper, which is "far better suited than glass to the Japanese house". Pervasive rather than invasive, and unlike the pierced light penetrating to the interior through the lattice screens of Arab cultures, diffused light depends on waxed-paper screens and shades to produce a soft illumination – one that makes the light almost tangible.

Not all paper is the same, however. The texture of Chinese and Japanese paper, for example, is ideal because it gives a certain feeling of warmth, of calm, and of repose. Tanizaki poetically points out that: "The same white could as well be one colour for Western paper and another for our own. Western paper turns away the light, while our paper seems to

Left Demetriov of Thessalonica in northern Greece sports a halo, just another name for a manifestation of prismatic colour, like a rainbow made almost tangible. In Christian art, an aura made visible as diffused light around the figure is known as a nimbus. Artificial-light manufacturers have been quick to appropriate both names. Halogen's radiant glow was so-called from the halo diffused through quartz and dichroic reflectors, while "Nimbus" (by the Outdoor Lighting Company) is a floor light that seeks to achieve that luminous effect by lighting from below.

Diffused

envelope it gently, to take it in like the soft surface of a first snowfall." Japanese-born American sculptor Isamu Noguchi used handmade paper to beautiful effect in the lanterns he made during the 1950s to diffuse electric light. Japanese paper is particularly suitable because it does not overheat, does not yellow, and it filters the light evenly. Tanizaki points out that the Japanese find it difficult to be really at home and to feel comfortable with objects that shine and glitter. "The Westerner uses silver and steel and nickel tableware polished to a fine brilliance but we object to the practice. We only begin to enjoy silver when the luster has worn off and it has taken on a dark, smoky patina." That was 30 years before halogen lighting was invented, the name for which comes from the halo of diffused light that appears like an aura around the head of saintly individuals in early Christian art.

Left Diffused light is seen when it is halfway between bright and dark – a hazy light in the woods, a sunset, a cloudy sky. Diffused light is not aggressive, it is the opposite of the drama of light, with chiaroscuro effect. Filtered through a forest glade, to no more than a presence, daylight is diffused in this 1902 painting of a pine forest by Austrian painter and designer Gustav Klimt.

Left The Pantheon in Rome, Italy, lit by the *oculus* in its dome, dates to the time of Hadrian's rule and was built 118–125. Until modern times, the dome was the largest ever built. The coffers originally had gilded-bronze rosettes in their centres. These were later removed to Constantinople. The Pantheon gave 15th-century dome builders inspiration that flowered in Brunelleschi's dome for the Florence cathedral in Italy.

Right When Le Corbusier beamed sunlight 20ft (6m) down shafts into the chapel at La Tourette, France, he painted the ceiling midnight blue and the inside of the three vast shafts white, orange, and blue, to colour the natural light. The pools of light can be almost physically felt.

Directional

Cities around the world always take into account the width that buildings need to be in order for natural light to penetrate their cores. Except in buildings intended for storage – such as warehouses, or in some types of factory and out-of-town supermarkets where light needs specifically to be excluded – the presence of natural light inside built spaces is of paramount importance. This is true even today, when modern artificial lighting systems could adequately illuminate internal spaces. Yet still we interrupt building façades with windows that are spaced with an age-old regularity that ignores the fact that the electric light has been invented.

Directional light can be thought of as the light that enters a building or focuses on the page you are reading. How daylight reaches the core of a building depends on the positions of its apertures – the domes, skylights, windows, or other openings. When light enters space in a rush from overhead, like a water course, it can be hard to distinguish the source because of the intensity of the illumination itself. If, however, the light entering is soft, like that from a bright but sunless sky, then space can be suffused with a more general illumination. How artificial light focuses on the page can have as much to do with the fitting as it does the light's beam width stated on its packaging. If the package states a beam width it is a spotlight, in which case the fitting will not change this. Fittings that control beam width have built-in reflectors and use ordinary light sources.

In the East, light is fragmented and split by lattices and screens that seek to exclude and channel it. In the West, domes and atriums are used as lighting devices to affect the penetration of light. The Pantheon in Rome gives a directional light that distorts the interior space and makes it appear larger than it actually is. The restricted nature of the beam increases the apparent distance to awesome effect, and this depth gives the light a diffused effect. Too much light impoverishes interiors.

Above The American architect I M Pei's glass pyramid at the entrance to the new wing of the Louvre in Paris, France, emits light. In order to make it more luminous, light is beamed from 10 recessed lamps by the German-based company Erco onto its steel structure, not the glass covering it.

Theatrical light

Rock concert sets are as celebrated as the stars. Musician and stage man Jean Michel Jarre (above) washes textures in sound and strobe light with laser beams. No sooner had the Berlin Wall collapsed in 1990 than British architects Mark Fisher and Jonathan Park rebuilt it in styrofoam (below) for Roger Water's (of Pink Floyd) one-night stand in the city. In "The Wall" – which was 550ft (168m) long – beams of light created a dynamic space, pin-pointing the darkness to create depth and atmosphere. Lighting designer Patrick Woodroffe says his biggest strength is darkness to play on the emotions with strobe lights and a flash of fireworks.

World's largest light sculpture

A neon sculpture called "The Sky is the Limit" above a moving walkway at O'Hare Airport, Chicago, Illinois, USA, is the largest light sculpture in the world. The 744-ft (227-m) kinetic sculpture by Californian artist Michael Hayden is composed of 466 coloured neon tubes programmed in sequences to provide continuously changing light patterns. The first and final sections of the three-section sculpture begin with white neon tubes to create a tranquil atmosphere. The centre section, however, reverberates with colour. The colours start at indigo and proceed through the spectrum, from shades of blue, green, yellow, and orange. The middle of the sculpture is bright red, and the colour pattern then repeats itself in reverse order starting at the mid-point. The walls of the terminal building have been built in wave patterns painted in complementary colours. The interior lighting causes the pebbled surface of the wall to glow in harmony with the sculpture.

Above American artist Bruce Nauman's *Seven Virtues and Seven Vices* uses LCD ticker-tape technology to convey an unexpectedly profound message. American Dan Flavin in his artwork made light both static and substantial, but Nauman keeps it energetically on the move spelling out a social commentary and using the medium of advertising slogans to reveal truths.

Right The great silent waves of the northern lights, which break the winter darkness of an inky sky, are an electrifying sight. The aurora borealis reminds us that light is indeed electric.

Above A short-lived wall of coloured light bursts into a blackened night sky – looking like welding sparks from a static source – except that here the kinetic light is carried aloft in a trajectory of fireworks.

Left Swarovski's cavern at their head-quarters in Austria is hung with facet-cut crystals and the experience is like stepping inside a glacier. It does not so much move as pulsate with light refracted from the zillions of faceted surfaces in an underground geodesic dome lit by fibre optics. A real crystal, examined from inside with the help of a laparoscope, was replicated by Swiss artist André Heller. Brian Eno's New Age techno-mix music plays inside. The entrance is found behind a waterfall that gushes from a mound, and a route takes you through the darkness to emerge in this magical chamber of moving light.

Kinetic

In 1893 American painter Bainbridge Bishop wrote: "The invention of electric light renders it possible to use color harmony as an accompaniment to a church organ and sacred music . . . beautiful effects could be produced by a combination of statuary and gauze curtains which, as the music pealed forth, would flash and fade with softly melting hues of colored lights with the chant of adoration." Light as a performing art was launched.

The phrase "being in the limelight" harks back to the days of gas lighting when lime was heated until it radiated. In 1840 the English scientist Charles Babbage devised a ballet including a largely abstract scene in which four coloured limelights would project a red, yellow, blue, and purple light that would move and overlap to produce a rainbow effect playing over the white-clad dancers.

Firelight and its successor for lighting, rushes dipped in resin, meant that light was always dynamic, never still. Over time, it changed until the neo-classical ideal of light was staid. We interpret light today as if it is the same every day, unchanging diurnally or seasonally, so that kinetic light at the end of the 20th century is associated with pop concerts and video art rather than something straightforward such as the changing light of day. French Impressionist Pierre Auguste Renoir painted the cathedral at Chartres, France, three times a day for a week to observe the changing light.

Its attention-grabbing qualities – from lightning to shooting stars – make kinetic light a favourite with artists. Lasers and strobes are the stuff of big spectaculars. Kinetic light is the raw material of fire eaters, and dance clubs nightly vibrate to it as tattoos made with light-sensitive ink illuminate the ravers. It can glare and then fade – qualities that make it interesting to architects and artists. In a video installation by American artist Bill Viola, watched in darkness, images of people on wrap-around screens emerge hazily and then, with a fizzle, burn out like snuffed candles.

Left Water is reflective. In Venice, the architecture adjoining this watery, mirrored surface has no paving to cut or muffle the light, and so it seems to float between two skies. The illumination inside the Venetian buildings at night can be extraordinarily beautiful because the watery bases on which they stand double the dramatic effect by reflecting and mirroring it.

Because we see most things around us – inanimate objects as well as other people – as a result of light from the sun being bounced from their surfaces, the physical world is perceived through the action of reflected light. Surface qualities either diffuse or intensify that light and thus change its hue and density, just as sunlight changes in the presence of clouds. Helping to explain why we see light as white, not coloured as it surely is, American Nobel-prizewinning physicist Steven Weinberg began *Dreams of a Final Theory* with an explanation of why chalk is white. Chalk is white because it reflects all the visible frequencies of the white light that illuminates it. It does, however, absorb infrared and ultraviolet radiation, but these particular frequencies of light are invisible to us anyway. All the other colours of light – from red and orange through to indigo and violet – are bounced straight back off the chalk, where they combine together to make the white we perceive the colour of chalk to be.

So why are only some frequencies absorbed by the chalk, leaving the others to be reflected into our eyes? The explanation lies in the interaction between light and matter. In the atmosphere, sunlight bounces off suspended water droplets, dust, particles of pollution, and so on and is scattered to form what we call fog. If this fog is dense enough, the light simply bounces around, getting nowhere. But when light – which is just a wave of electromagnetic energy – cannot complete even a single cycle of oscillation before being obstructed, it stops dead in its tracks. Light frozen between an alloy of two metals, gallium and arsenic, is crucial to photo-tonics, in which computers use photons of light to do their stuff more rapidly than is possible using a stream of electrons.

Reflective

Left Natural and artificial light are balanced in an interplay of surface and depth in this detail of a picture by French artist Claude Monet of the waterlilies at Giverney, France. The sun lights the surface and the shadow, which comes from the planting, gives form, depth, and substance to the painting.

The reflective qualities of light preoccupy lighting designers just as much as they do physicists. Photosensitive 3M plastic film, made in sterile laboratory conditions in Texas, USA, allows for no diffusion of light at all. The light from any source, channelled within a tube, travels through the film where it is distributed completely evenly. This is why the material is so effective for highway road signs. As well, light can be sent a long way from its source and so there is no heat generated.

Leonardo da Vinci's *Last Supper*, perhaps one of the most famous works of the Renaissance, painted in 1497, can be seen in Santa Maria delle Grazie, Milan, Italy, where it is illuminated by light emitted from 3M photosensitive film in a cylinder that runs along the base of the recently restored fresco. The absence of heat generated by this light source is vital for the long-term preservation of the colours of the painting and the integrity of the material surface of the wall itself.

Left The hair of the polar bear acts like a fibre-optic cable transmitting reflected light from the snow. Looked at under a microscope, a white polar bear's hair is, in fact, completely transparent, and it conducts light down from its snowy habitat to the animal's black skin. Even the pads of its feet are furry in order to reflect the white light it needs for effective camouflage.

Above Seen from the river, American architect Frank Gehry's Guggenheim Museum in Bilbao, Spain, reveals an unexpected familiarity with its surroundings through reflections on its cladding. Titanium blends with tensile strength to accommodate Gehry's implosive geometry, in which conventional pillars and beams disappear.

Right With its vivid tonal contrasts, deep shadows adjacent to bright highlights, rather than bland, overall colour, *Metropolis*, by Fritz Lang (1926), is a film famous for its luminous effects. In the huge futuristic city of Metropolis an army of slaves, robbed of their humanity, works underground as part of gigantic machines, when an out-of-control robot starts a revolt.

Far right Steven Spielberg's *Close Encounters of the Third Kind* (1977) observes a formal geometry in the processional routes and grid of lights. UFOs fly over the United States in collusion with the government, which broadcasts musical notes linked to flashing coloured lights in an attempt to communicate with them. The original *son et lumière* show has special effects that are more rational and modern than those found in most sci-fi.

Right Stanley Kubrick brings to chilling life HAL, a computer in charge of a manned spaceship in *2001: A Space Odyssey* (1968). Amid the blinking lights and fluorescent panel displays, HAL loses its grip on reality as its programming leads it to the conclusion that the human cargo is a danger to the mission. This was the first film in which most of the action was shot inside a spaceship while light levels were kept low.

The intensity of the experience of entering or leaving a field of light is celebrated nightly in theatres all around the world. Just before the performance is due to commence, the front-of-house lights fade to give way to a rapidly deepening darkness. A hush falls over a restless audience. Anticipation builds. The curtains rise. The stage lights snap on. This is the drama of light.

Outside the theatrical world of fantasy and the author's imagination, on the daily stage of life, light still retains its ability to surprise and transform, to make-over the mundane, prosaic, and everyday experiences. If we continue to throw too much illumination at our bland, evenly lit, and homogenized interiors, we are in danger of losing the ability to really see. Blinded by too much light, we easily become bored by the flatness of an unchangingly luminous environment. Often, what can only be partially seen has more of an air of mystery and drama than that which is clinically revealed and, therefore, will intrigue and stimulate the eye, holding our interest for far longer.

Through the medium of film we discover that light cultures differ from country to country – and that these cultures are as clearly differentiated as the rising of the sun in the east and its setting in the west. Filmmaking grew up with the invention of artificial lighting systems and it is the manipulation of light that represents the drama of the experience. We have become so educated in the methods of the lighting director that we instinctively equate different degrees of lighting with different degrees of dramatic content. You are more likely to be set on the edge of your seat by what you cannot properly see, but can well imagine.

The drama of light

Eastern scenography

The films *Raise the Red Lantern* (left) by filmmaker Zhang Yimou and *Farewell My Concubine* (above) by Chen Kaige are two successes of the 20th century from China. They give an insider's view into that ancient territory through the people and the places. Porcelain-white faces, rosewood-dark fretted lattice, sensuous, brilliantly coloured silks, and white clouds of smoke somehow resolve their differences under the smouldering joss sticks or cherry blossoms, as evocative as light fragrant from firewood. Light seeps from the lanterns and under the doors – a diffused and warm light that reflects the age-old mystery that is China. Though the imagery of Zhang and Chen is stylized, and may even be historically inaccurate, the light pictures they paint are hauntingly beautiful.

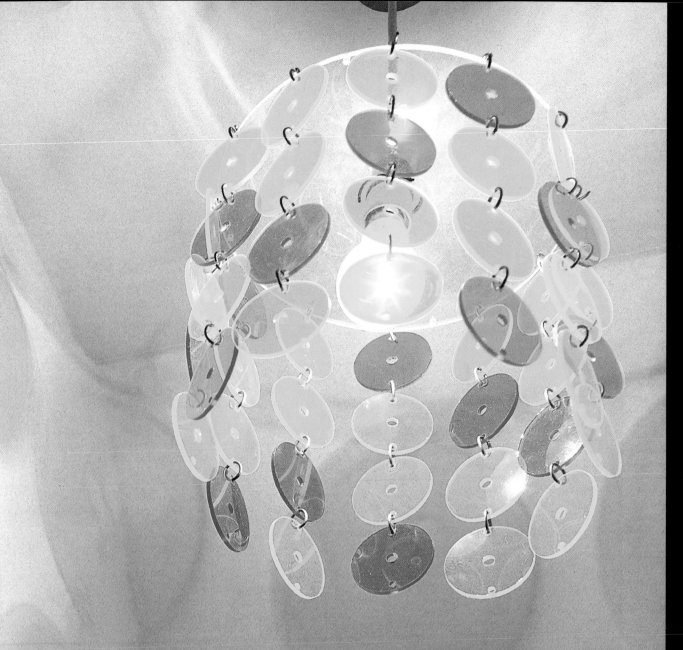

Right "Costanza", with its cone shade, square base, and narrow rod stem, is a classic reinterpreted through light-sensitive material by Paolo Rizzatto for Luceplan. Aircraft cockpit-panel material called bayern evenly diffuses the light from its core. The proportions are variable because of its telescopic stem. Powered by the Edison E27 bulb (lamp), the output can be dimmed by tapping a touch-sensitive dimmer.

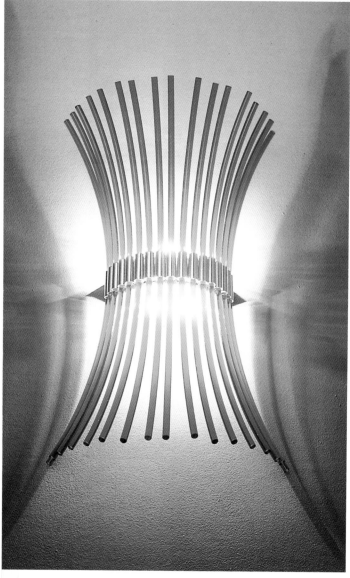

Left Multicoloured-glass rods bundled into a steel band both filter and colour the light from a clear incandescent source in "Tamiri Parete" by Roberto Pamio for Artemide, available as either a hanging or wall light.

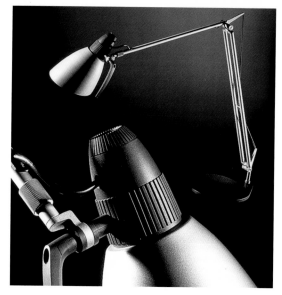

Right Another dual-identity task light "Logos" by Martin Huwiler (1998) for Belux AG takes both compact fluorescent and incandescent bulbs (lamps).

Below left "Dom Piccola" ceiling light by Roberto and Ludovica Palomba (1998) for Foscarini, uses either incandescent or compact fluorescent sources.

Right Maestro Achille Castiglioni in his 85th year (1998) telescopes his "Diablo" pendant for Flos to pull down an incandescent light with a cone for indirect illumination. More architectural than decorative, the fluid pendant hides the mechanism in the inverted cone.

Left The ultimate diffused light is the "Glo-Ball" for Flos, designed by Jasper Morrison (1998), which is a flattened globe in opalescent glass. When lit up, the globe looks curiously two-dimensional, an effect that is heightened by the fact that it casts absolutely no shadow. Overcoming the problem of how to support such a weighty piece of glass on such a slender stem, "Glo-Ball" is an example of thoughtful design applied to shade and diffuse the light coming from a familiar old incandescent source. Morrison argues that the quality of the light is the most important thing. You can't see the light source directly and, at the same time, it is not obscured.

Right "Aura" by Hannes Wettstein from Belux AG is essentially a bulkhead light design. It can accept a coloured rim as well as both incandescent and energy-saving bulbs (lamps).

Above Low-voltage allows lighting designer Ingo Maurer to bring light control literally to your fingertips with "Symh-oh-nia", a further development of "YaYaHo". To run the system, standard current has to be reduced to 24v by a transformer fitted with an integrated "Watch Tronic" power-monitoring system and a thermal fuse.

Right The star of the starburst ceiling is the dichroic lamp, such as "Tijuca". Recessed halogen downlighters marked a turning point in the history of lighting as they became architectural installations, not freestanding fittings. In-built transformers were hidden in the ceilings, and lights had to be positioned and cabled before plastering. For many centuries, light has been associated with warmth. Now, halogen is shining on a generation growing up under a cooler light.

Left London's first pontoon bridge, by Future Systems, glows as luminous green as a caterpillar as it stretches across the Thames at Canary Wharf. Lights underneath the bridge illuminate it by night, and all along the wooden deck halogen floor lights inset in regularly spaced lines on either side change colour as the walker approaches. The halogen light itself does not alter, just the dichroic filters housed in the large halogen lamps, which break the light into prisms of rainbow hues depending on your angle of approach. Colours change from pink to gold, green, blue, and violet.

Halogen and dichroic filters

One step on from incandescent technology and a pocket of halogen gas inside the glass envelope prevents the tungsten filaments from over-heating, vapourizing, and finally settling on the inside of the glass walls to darken the light output. Smaller filaments in a more stable environment need smaller envelopes, so the introduction of halogen started the trend to downsizing. At the beginning there were some technical glitches – for example, halogen contained at high temperatures simply evaporated. To overcome this, quartz replaced the glass to cool things down. Tungsten-halogen is energy-efficient, has a long life, and remains bright-white throughout its life, but it does require a transformer to deliver the low voltage it uses. As well, it heats up when switched on. Dichroic filters, which are a further development of halogen, maintain the same amount of light as regular halogen but direct infrared heat behind the lamp, which works at a lower temperature. Incandescent light emits in all directions, whereas halogen-quartz with metal reflectors directs the light where it is needed – another reason for its popularity with lighting designers and architects.

The advent of tungsten-halogen light in the 1970s had an impact on 20th-century architecture. It marked the moment when decorators handed over control of modern lighting installation to architects. The "Tizio" table light, by Richard Sapper for Artemide in 1973, was the first design driven by the downsized halogen lamp. It removes the need for wiring with a positive and negative circuit supported in the counterbalancing weights.

Halogen changes the way we see things. In the cave, firelight was warm. Gas light was warm; so is incandescent. We are still influenced by this evolutionary memory, and in our homes we use lights that replicate the effect. Halogen, with its cooler output, is changing all this.

Left An architectural use of halogen can be seen here, this time installed in the fabric of the walls to illuminate the treads in a stairwell. Halogen's longer-lasting lamp life made it popular for public spaces, but compact fluorescents have outstripped it in terms of energy-saving performance.

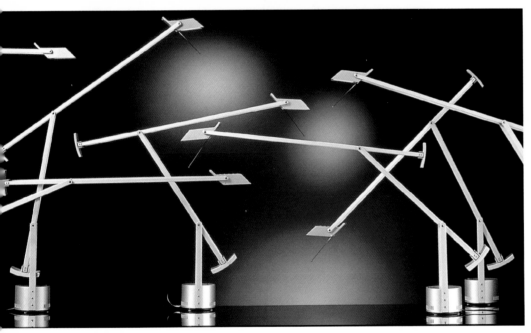

Left Pattern-making with light and shade is the intriguing result from the striated "Fida", a wall fitting from Artemide by Mario Botta, architect of the San Francisco Museum of Modern Art, who uses a screen to regulate indirect light emissions. An adjustable head in a white-painted perforated plate takes two 12v halogen lamps to beam fine lines onto the wall.

Left Ground-breaking "Tizio" by Richard Sapper for Artemide has a small halogen lamp in a reflector, and the jointed arms with counterweights power it through positive and negative terminals – the world's first wire-less light. Only halogen lamps were light enough to execute this balancing act with the delicate flexible arms.

Above Blown-layered Venetian glass, shaped like a galleon in full sail, by Valerio Bottin (1998) for Foscarini illustrates how a translucent glass shade acting as a diffusef can increase the volume of light. Halogen is a compact light source with a wide beam emission, and harnessed in this way it makes a dramatic pendant.

Below Water flasks like these are everyday objects in Spain, so Josep Lluscà turned one into a light, the "Bolonia", by Metalarte in Spain, which diffuses light from a halogen lamp within the glass container. As well, he turned it into a task light with an adjustable shade, which is the stopper made from opalescent glass diffuser.

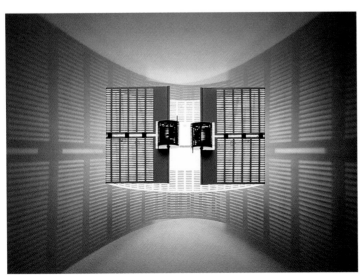

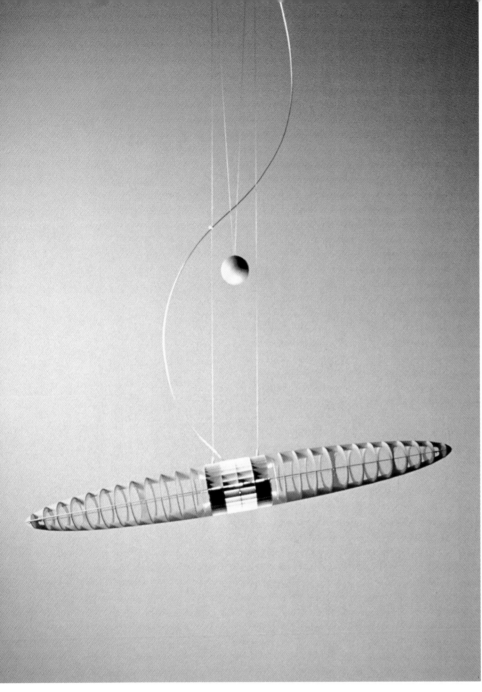

Right The two Es in "Eedisson" stand for Energy management and Emotion, the two Ss for Safety and Saving. Edison's original incandescent is translated into a 12v halogen lamp operated through a transformer with an energy-management function by Belux AG. Urs and Carmen Greutmann designed it in recyclable plastic.

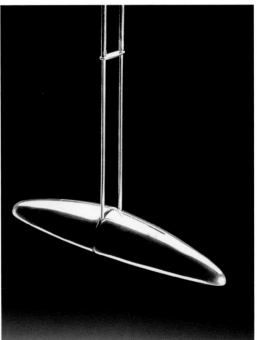

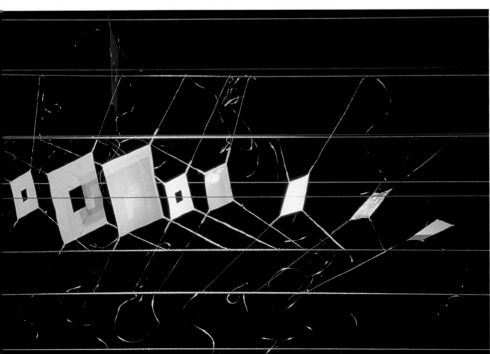

Left "Symh-oh-nia" is the new addition to the trapeze-like system of lights by Ingo Maurer. All the wires and parts of the fitting, as well as the small birdlike, fanciful halogen dichroic lamps on wires attached to transformers on the wall, give the impression of looking at the night sky. They literally turn light into energy as they fly across their wires when switched on. This installation is at the Louisiana Museum of Modern Art, Denmark.

Above left "Titania" is an identifiable flying object by Paolo Rizzatto and Alberto Meda for Luceplan, which comes with different-coloured interchangeable poly-carbonate filters to fit its elliptical form. Yet the light itself is not coloured. White light from a halogen lamp, positioned to beam directly downward, strikes whatever is beneath it, while only weak light is emitted from the sides. Colour shows only on the body of the fitting.

Above To celebrate the Barcelona Olympics of 1988 Jorge Pensi designed the gleaming "Olympia" for Belux AG to look like a projectile missile. Two halogen lamps are housed in its elliptical aluminium body, which is suspended by a pair of stainless-steel rods. In the same family of lights there are freestanding wall and table lights as well as a wall-mounted downlighter.

Above The French designer Zebulon presents the clean lines of "Buro" for Artemide. Built-in electronics allow the temperature of the light to remain stable, eliminate flicker, give longer light duration, and minimize the overall weight of the wall fitting.

Right HID lamps produce light by passing an electric current through gas or vapour under pressure in a glass envelope. Metal-halide (near right) and sodium (far right) have replaced old mercury-vapour types.

Above The "M.I.L." wall-mounted uplighter from Concord has an adjustable blind that gives a soft-edged ceiling cut-off line. Screen-printed safety glass improves light distribution.

Below The white glow of metal-halide is evident in this freestanding room light called "Spoon", by Georges Adatte, Etienne Ruffieux, and Eric Giroud (1996) for Belux AG.

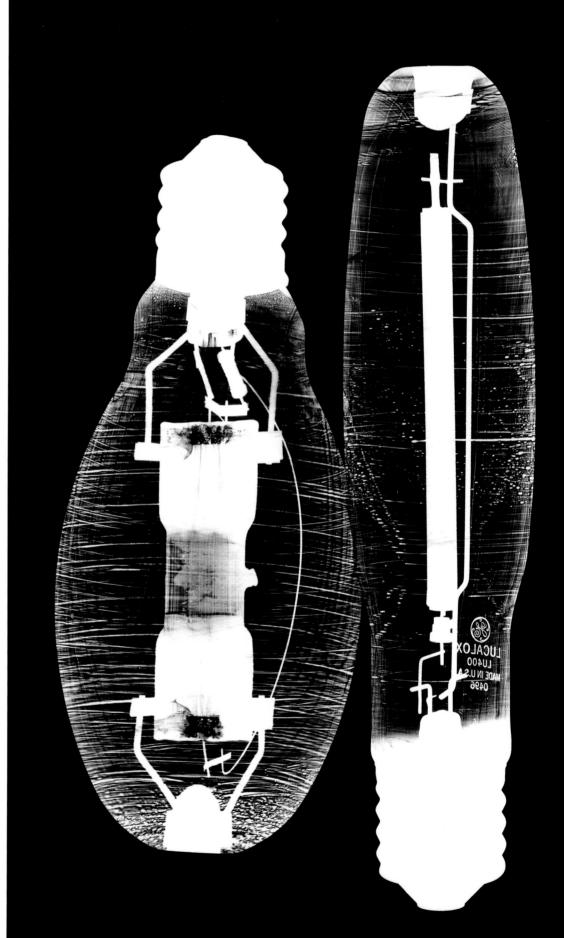

Above A series of Philips metal-halide discharge lamps.

Right Street lighting's eerie glow from old mercury-vapour lamps, first introduced to New York in 1901, has been replaced by high-pressure sodium. The Brooklyn Bridge has never looked better. Metal-halide, with its more accurate colour rendering and compact size, has also moved indoors.

Metal–halide and sodium

The World Cup 1998, which used venues such as the impressive Stade de France stadium, was played under the white illumination of metal-halide lamps. Metal-halide, which came along in the 1970s, is the latest version of what are known as high-intensity discharge (HID) lamps. HID is the generic name for a family of lighting types that are more lumen-efficient than incandescent or fluorescent. Like fluorescent, they produce light by passing an electric current through a gas or vapour held under high pressure. HID lamps were first used for street lighting in America in 1901. These used mercury vapour, but many US cities have subsequently banned them in an attempt to minimize light pollution of the night sky. On clear nights, the Milky Way can now be seen once more over downtown Tucson, and in 1997 New Mexico declared itself a dark-sky state with regulations on businesses using after-dark lighting and a total ban on mercury-vapour illumination. Streets are no longer being flattened by a

sickly bluish glow, or the peachy-yellow of low-pressure sodium, which is now mostly used for warehouses or roadways. Metal-halide is the newcomer that does not flatten or drain colour from streets or stadiums. Energy-efficient, downsized metal-halide lamps have moved indoors, too.

Discharge lamps need literally to generate a discharge between the two ends of a tube. It is not easy for the flow of electricity to move from one electrode to the other, so it needs a pulse of power to push it along. This is why each HID tube has a starter switch inside of it, along with a ballast that stabilizes the variable surges in power and makes it flicker-free. The electric discharge in the gaseous atmosphere within the tube generates invisible ultraviolet rays. The white part of the glass tube is coated in a phosphorous material that is sensitive to the ultraviolet and acts like a transformer, absorbing the rays and re-emitting them as a longer wave-length. So it acts not only as a diffuser but also as a producer of light.

Right The fluorescent light – introduced widely in 1938 for advertising billboards, came bottled in a long tube and was known as neon – has evolved into these new compact fluorescents. They use the same technology but add a ballast for flicker-free, long-life, low-energy lighting. Special coatings on the phosphorescent glass help to minimize the cool white light of fluorescent, though there are lighting designers, such as Toshiyuki Kita, who will not use it.

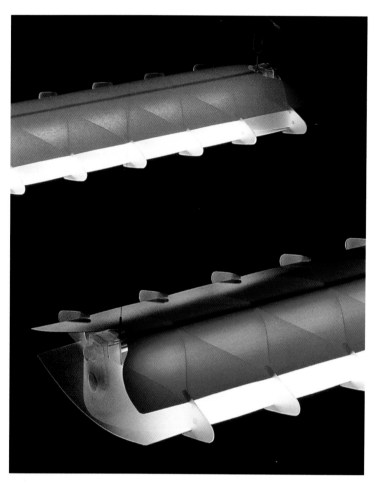

Left "Bandoneon" refers back to former times of colourful neon light displays in this wall fitting for Belux AG by A Aebi, A Louis and P Reymond (1998).

Right Dan Flavin created works of art using colourful industrial fluorescent tubes, which he wedged in corners, stacked in columns, and bunched or arranged in barriers to show how light can define space. His piece in an art dealer's London apartment, designed by architect John Pawson, uses tubes of strip lighting to disrupt the actual space of the room as the lights wane and shadows are created.

Fluorescent

In 1938, when advertising billboards went electric, fluorescent was known as neon. Once large and ungainly, the new compact fluorescent that folds back on itself in U-bends is now comparable in size and weight to incandescent bulbs (lamps). They last an average of 12,000 hours, which is about 12 times longer than an ordinary light. By comparison, halogen lamps last between 2000 and 3000 hours. Years can pass before you need to change one, and this is why the Osram Dulux EL won an environmental award in 1996 from the European Union.

The reason you never hear the term "neon" anymore, even though cities around the world blaze with it at night, is because in the early days voltage fluctuations meant that tubes flickered and then shattered. Power supplies can fluctuate as much as 50 times a second, causing lamps to switch on and off. Thermal inertia in halogen and incandescent light sources does not allow the filaments to cool down and so no problems result. However, fluorescents suffer from discontinuity, which causes them to flicker. Even if our eyes don't register the oscillations, our brains do.

Modern compacts have a built-in ballast to regulate current. They are also colour corrected, with coatings in a range of colour temperatures from warm yellow to cool blue. At the Luceplan headquarters outside Milan, Italy, light boxes lined up along the corridor launched an architectural lighting system called "Orchestra", using a new, slim-line compact fluorescent with aluminium or gold reflectors to project light out and up. Because it gives a neutral light output you can repeat it in any configuration, and "orchestrate your own show" in the words of the light's designer, Paolo Rizzatto.

Fluorescent light visually solidifies in its tube, like a *Star Wars* tool where solid light becomes a defence. This notion of defence through a light beam was explored brilliantly by the American artist Dan Flavin.

Left Fluorescent comes full circle with "Cyrcline" by Jürgen Medebach (1994) for Belux AG, which has all the graphic impact of a logo. The universal eco-light, with an energy-saving fluorescent tube, is not pointlessly covered with glass or with an added reflector, but the light output is fully utilized.

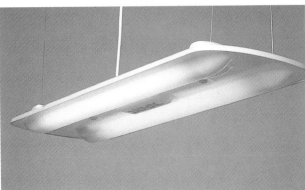

Left Light as an accent detail as the manta-ray-shaped light fitting "Dumoffice" moves on tracks in the home or office. The armature is moulded Plexiglas, which enhances the glow, and the light comes from two fluorescent tubes, which spill colour from around the edges like a colourful trim.

Left Post Design is where the Memphis movement ended up. "Art 09" by Johanne Grawunder (1997) is an industrial interpretation of the Dan Flavin approach to fluorescent, dissolving the space around the spill of light. These tubes deny access to the recesses behind them, which appear to be filled with light.

Left The light "Juliett Victor" by Dietrich Brennenstuhl and Sebastian Mohn for Nimbus GmbH recalls the old days of air waves. The light appears to emanate from the fitting in a series of waves.

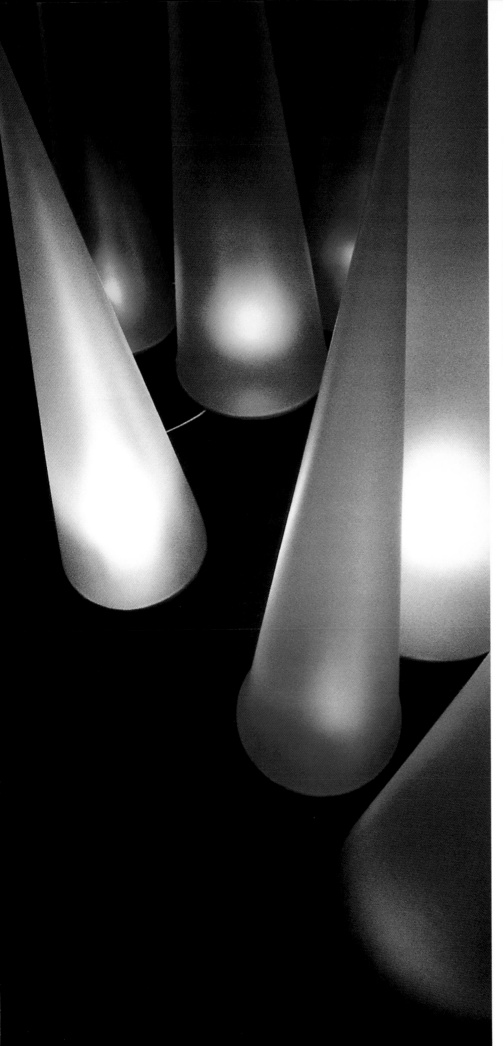

Left "Inflatable Cone" lights in a cluster were originally part of an installation at the Royal Institute of British Architects. Katrien Van Liefferinge bridges the gap between art, architecture, and appliances for the home with these playful lights utilizing a compact fluorescent light source. She calls them "light objects", and likes to emphasize the crossover between art and functional design, but stylists have presented them as post-modern Christmas trees.

Above "Economy" is an energy-saving system by Jürgen Medebach (1991) for Belux AG that provides mild direct and indirect light for living spaces and work areas. Four compact fluorescent lights approximate the luminous power of 10 standard incandescent bulbs (lamps). A decorative strip for the light fitting provides a splash of colour.

Below "Boxer" is the tough, don't-mess-with-me name for a heavy-duty work light by Konstantin Grcic, for Flos (1998). The die-cast brushed-aluminium or black holder accepts fluorescent tubes with a series of different diffusers. In addition to being technically well designed and architectural in appearance, it is also a robust, anonymous light for what Grcic calls "the forgotten areas of cellars and garages", to bounce light off the wall or reflect it off the floor.

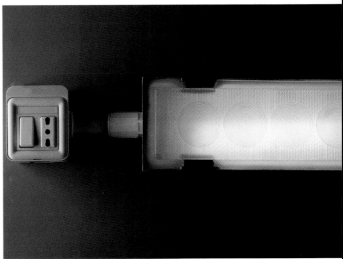

Left Light can loop and travel in curves along fibre-optic cables. To prove it, Athena created a shimmering curtain wall of light at the Cologne Furniture Fair 1998, demonstrating the fluid way in which light can be made to perform.

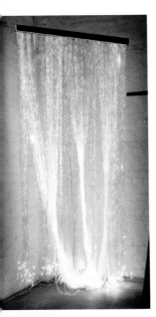

Right A web of optical fibres by Javier Mariscal spreads across a clear rotating canopy in the Sun and Moon nightclub in Barcelona, Spain. Entering the Torre d'Avila nightclub in the former Expo tower, built in Barcelona in 1929, was like stepping inside a lava lamp. Sadly, it closed at the beginning of the 1990s.

Left "Kika", by Maurizio Caimi, is a bookcase with integrated fibre-optic lighting. The wood is natural beech with special glass panes and 20 points of lighting provided by 6mm optical-glass fibres.

Left Josef Hoffman designed this button-back chair in 1911. Shiro Kuramata in his one-off "Homage to Josef Hoffman Vol 2" replaced traditional upholstery nails with pinprick fibre-optic light, operated and adjusted by remote control. They literally lighten its solid form while picking out the flair of the design.

Right A fibre-optics "Octopus" generator lighting system by Philips, which can be used for downlighting, spotlighting, logos, or colour projection.

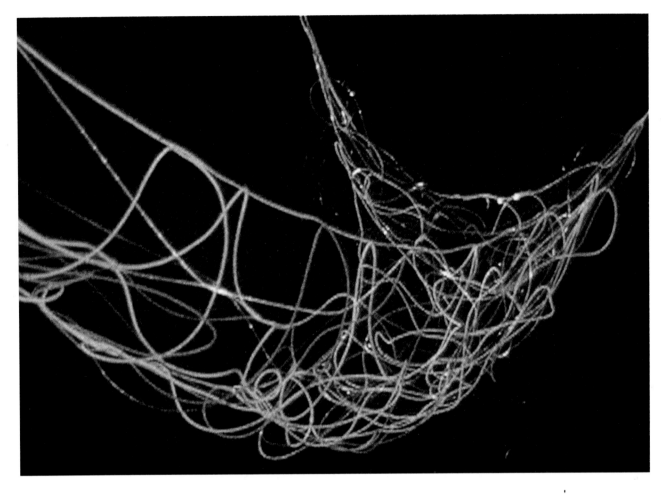

Left At Canary Wharf in London's Docklands, installation artist Diana Edmunds strung a net of woven fibre-optic cables tethered to yachting wire across the water. When the installation, "Grid", moved to a site in Wales she allowed the water to wash over it to enhance the light's ebb and flow. The light source for the fibre optics housed in a light box was sufficiently far away to allow the installation to be immersed in water. While the structure of the work is important to the artist, the light is the paramount ingredient. Edmunds likes the quality of the light, which can approximate moonlight or sunlight depending on whether she uses a halogen or a fluorescent source in the light box.

Fibre optics

You need a ruler to plot the trajectory of light. In normal circumstances light travels in straight lines. Yet via a fibre-optic cable light can be conducted along a curved path – as can the electromagnetic waves used for carrying telephone messages. Bundles of very thin, even hairline-fine glass fibres transport light over great distances, either allowing it to emerge all along their pathways as wavy rods of light or confining the light *en route* so that it is released only at the tips of each fibre to form a glow-worm-like cluster. Fibre-optic cables can bend, and be woven, which is why the lighting installation artist Diana Edmunds likes them so much. Her "Light Reed" in a lake at the James Dyson vacuum-cleaner factory in the English countryside reveals itself only as darkness falls, when the light glows from the base of 150 rods. Invisible until the night deepens and shadows move across the surface of the water, the cables bend and sway in the wind just like reeds.

It is difficult not to be impressed by pinpricks of bright light illuminating a Rolex Mariner's watch at the bottom of a glass tank full of water and fish. Travelling along the fibre optics, halogen light from a box some distance away cools down. Since it is remote from its power source the light can be safely submerged. Fibre optics are the coolest lights around. All you need is a transformer, a light box, optical fibres, and a light outlet element. The smallest light box is about the size of a shoebox. The bigger the box, the greater the number of fibres it can power – up to 350 – because the box has to accommodate a cooling fan.

Light conducted by fibre-optic cables is dust-free, does not contain any ultraviolet rays, does not generate heat, and has no visible lighting elements because the light box is hidden. Colour fidelity is good, too. In museums and showcases, objects can be clearly lit in an enclosed space without heat and changes in the microclimate.

Left Manufactured in a Texan Laboratory, 3M plastic film, normally used to transmit light from a metal-halide source far away to illuminate road signs, has become a luminous "Virtuel" by French architect Jean Nouvel for Luceplan. 3M film does not allow light to seep out of the sides, but reflects it back so that there is little illumination wasted. This even, glowing, luminous light – "like a column of mercury" says Nouvel – is more about the new technology of light-emitting material than the actual shape of the light.

Above Click a button on a remote control and this light, the "Diafani", will give you a sunset, dusk, or bright noon-day light. Its designer, Ernesto Gismondi, calls it "a projector of sensations". Artemide's "Metamorfosi" programme uses polychromatic light systems linked to remote controls so that the user can mix different colours to create a variety of tones and shades that respond to diurnal changes. The "Diafani" can be mounted on walls or ceilings and its electronic regulator and remote-control sensor are built into the support base.

Right The light for your terrace or garden is high, low, or stuck in the ground. It can be hung individually or in clusters looped, for example, over the branch of a tree. Ross Lovegrove designed "Pod Lens" for Luceplan to take an energy-saving E27 electronic bulb (lamp) or an incandescent light source. The light cable is the same colour as the injection-moulded polycarbonate with prismatic sides used for the body of the light. The material is self-extinguishing and the colours are ultraviolet resistant. The fitting is also resistant to humidity, water, and snow.

Right "Solar Bud" by Ross Lovegrove for Luceplan collects solar energy in a photovoltaic cell that powers rechargeable nickel-cadmium batteries. When the sun sets, a circuit turns on three red LEDs. Left in daylight for four hours it gives 14 hours of light.

Right Dichroic lamps in a pendant fitting, by Antonio Citterio and Glen Oliver Löw, control, point by point, the surface to be lit. This flexibility of lighting effect is provided via a printed circuit board incorporated on the upper surface of the sheet of glass. The lamps are connected one to another by a low-tension circuit.

Lights of the future

In the 21st century illumination will come in the form of light-emitting material, rather than being sculpted and moulded from a material that conceals the light source. This exciting technological breakthrough begins with experiments into new materials. Take, for example, holograms that are bright enough for you to read by suspended in the air like a ghostly pear-shaped incandescent bulb (lamp), which its designer, Ingo Maurer, calls "Edison, where are you now that we need you?" Maurer predicts that liquid-crystal displays (LCDs) – the same sparkly light source you see as auto brake lights – will be an important light source of the future.

How light travels from its source to its destination is crucial to designers. Once channelled along fibre-optic cables – which consist of bundles of very thin, flexible glass fibres – light, with its tendency to travel only in straight lines, can be persuaded to follow any shaped path that is required. 3M plastic film, made under special laboratory conditions, also carries light long distances from a remote source to illuminate anything from road signs to works of art. This could not be possible if it were not for the fact that the film does not allow light to scatter or seep away, and that the heat generated is left back at the source, where it does no harm. Remote-source lighting is an important new tool for designers. The film works like a magnifying lens – instead of the light dissipating as it travels along the cylinder, the plastic film stops the light in its tracks and reflects it back on itself as if it were a mirror. Its white light is curiously luminescent, which is why French architect Jean Nouvel harnessed it in a freestanding cylindrical light for Luceplan, "Virtuel".

As fossil fuels run out and global warming is underway, designers are tapping into the sun, our main light source. Ross Lovegrove has designed an outdoor torch in the shape of a giant golf tee, which stores sunlight during the day and releases it at night.

Right "Ra", with its overhead sun-bed styling, is the first system to marshal into one channel different wiring elements and connect them to the power grid with signal and control lines built into the devices linked to it. Telephones, light, fire and smoke detectors, alarms, air conditioning, and public-address systems are all integrated in the "Ra" system, designed by Sir Norman Foster for Artemide.

Below Controls from Italian company B-Ticino bank thermostats, infrared detectors, sound-diffusion systems, and dimmer switches on modules covered in materials that have a feel-good factor – rubberized plastic, galvanized steel, graphite paint, and transparent meta-crylate. The company believes that switches represent that magical moment of going from darkness into light.

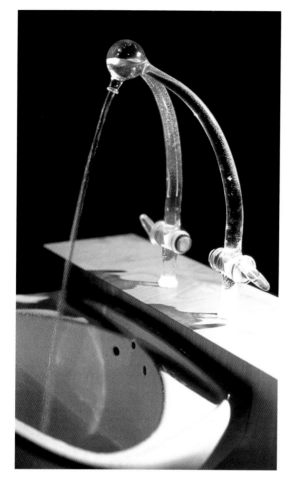

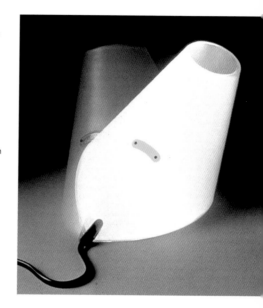

Left Designer Arnout Visser for Droog (which means "dry" in Dutch) uses light to make temperature visible with the "dd49" glass taps or faucets that channel red light along the hot-water pipe and blue light along the cold.

Above "Blow", by Ferdi Giardini for Luceplan, is a fan-light that has blades as weightless as a dragonfly's wings, coloured or transparent. The fan – and the light, which uses either a compact fluorescent or a halogen lamp – can be switched on from an infrared remote control, or by a switch on the wall or hanging from a wire. The direction of the rotating blades can be reversed to stop that stale, nondirectional flow of air.

Right The "On/Off", designed by A Meda, F Raggi, and D Santachiara (1988) for Luceplan, demonstrates that freedom from switches is possible right now. On many lights switches are already obsolete, with touch-sensitive fingertip dimmer controls replacing clunky switches on cords. With "On/Off", the whole unit switches on or off at the tap of a fingertip.

Controls

"When there is nothing in a room, the exact position of the light switch requires the most careful consideration if it is not to become a discordant element, disrupting the proportions of the wall which it interrupts." So believes British minimalist architect John Pawson. Never mind the location of the switches, just look at their archaic form. In the same way as the performance of autos is still judged on "horsepower", so too do switches hark back to the turn of the century, when Thomas Edison piped electricity into New York using gas pipes to carry the cables and distribute the power. Switches developed like triggers to shut down the gas supply with an audible clunk were retained for electricity, even though the switch was never connected to the supply – the obvious switch system we are all so accustomed to was obsolete even when it was pioneered.

As the world shrinks in the slipstream of jet travel, simply recharging a laptop requires multiple adaptors. Every country in the Americas and Europe differs. The domination of world power systems is the ultimate control freak's goal. The prototype may be installed in a house in the USA belonging to the world's richest person, Microsoft's Bill Gates. He has cabled his house to respond to anyone wearing a special electronic pin. When it is dark, moving zones of light accompany them as they walk about. Entering a room causes the lights to come on, while the vacated room dims. If the telephone rings, only the nearest handset registers. The pin allows the house to identify and locate the inhabitants and it accepts instructions. It tells the monitors in each room to become visible and chooses images to beam out from the 40-in (100-cm) screen walls stacked four high and six across. Gates wanted the screens when not in use to vanish by displaying on them a wood-grain pattern matching the panelled surroundings. However, a monitor emits light while real wood reflects it. So, he compromised by sliding them away behind real wood panels instead.

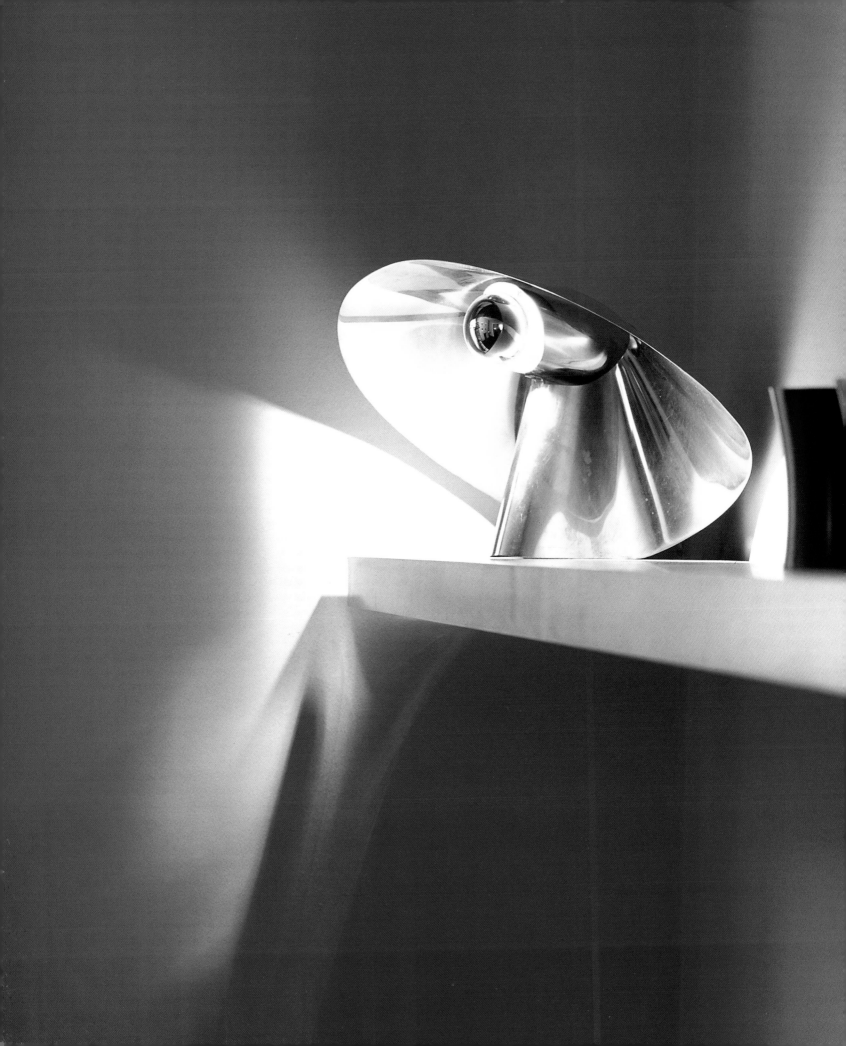

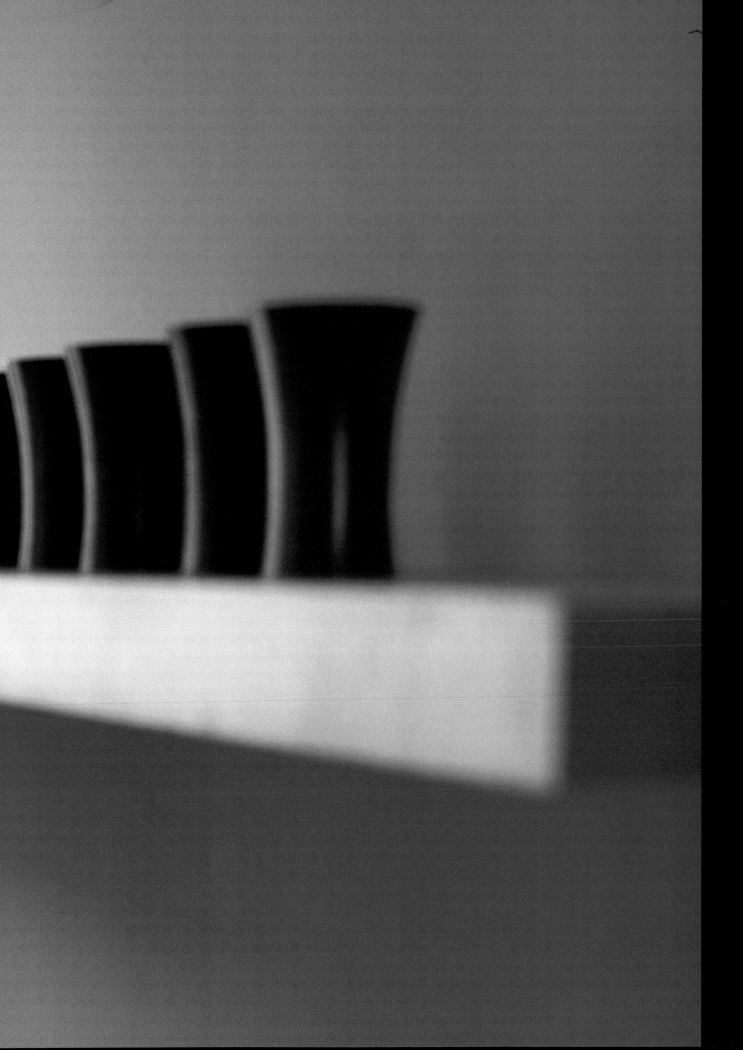

APPLICATIONS

Learning from Soane

Long before the invention of electric light, the lighting genius of the English architect Sir John Soane (1753–1837) controlled the quality of daylight through the use of domes and skylights fitted with coloured glass to warm the gray, northerly daylight and with reflective surfaces to bounce back the light. The atmospheric effects of light and space he created can still be observed in his tall, narrow, terraced house at 13 Lincoln's Inn Fields, London, which is now a museum. Soane urged his architectural students to "discover the many effects of light and shade which close observation of nature alone can give". Nature remained the great model, too, for Soane's fishing partner – JMW Turner.

Whatever the weather, inside it is sunny. Yellow glass in long skylights outside the picture room at 13 Lincoln's Inn Fields takes the deathly pallor off white plaster casts set along the corridor. There is also a sunny glow in the breakfast parlour from the aperture in a dome at its core. In the monks' parlour – as the retreat for afternoon tea was called – Soane sought another emotive lighting experience by using dark colours in the restricted space with the cathedral-like effect of stained glass playing over a scrubbed-deal floor.

Soane ingeniously used mirrors to expand the illusion of space, placing them behind exhibits and above bookcases to suggest rooms beyond. A Canaletto drawing, framed and hung in Soane's dressing room, is mirrored on the back, which implies that at some time the frame was reversed so that the back became visible. Sheets of mirrored glass fitted in front of the pier table in the house's library reflect the Axminster carpet in a calculated effect to push out the physical boundaries of the room. Every alcove is mirrored and there is glass set into the floor and on either side of the fireplace to bounce back more light. At night, the rooms sparkle in the candlelight and lamp light.

Far left The Soane family lived at 13 Lincoln's Inn Fields from 1813 until Sir John's death in 1837. Their breakfast parlour survives intact to this day in the museum. Light from overhead glazing set in a narrow aperture defines space. Walls at the pendentive dome's perimeter have disappeared so that the room is redefined by screen walls beyond. Mirrors inset around the domed ceiling bounce this focused beam of natural light all around the far corners of the room.

Left Painted in that agreeable Georgian stone colour known as "string", the central corridors of the museum outside the picture gallery are filled with busts. Presiding over the space is a bust of Sir John Soane himself. Yellow glass inset in long skylights bathes the chalky white plaster figures in a welcoming warm glow. Deal floorboards are scrubbed white to reflect light upward.

Right The amber glass bathes the casts from an antique statue of the Apollo Belvedere in a golden glow. The glass panels are set in the pendentive dome at the core of Sir John Soane's house in London. Soane designed the pedestal himself.

David Hicks

Flamboyant, opinionated, at times outspoken, David Hicks (1930–98) pursued the elegant and beautiful in interiors with a single-minded intensity. Sir John Soane was one of his heroes – "a genius and also a cross old man like me," Hicks admitted in his final, 68th year. He adapted some Soanian ideas on lighting for his own apartment in London, creating paths and destinations by controlling beams of directional light and varying light sources. Like Soane, Hicks bathed his space in a warm glow, using incandescent light rather than candlelight and amber glass. He abhorred white fluorescent directional lighting. Apart from incandescent table lights and wall sconces, all other sources of light in his apartment were hidden, either eyeball spot downlighters recessed in the ceiling or freestanding halogen uplighters.

Daylight is harder to control than electric light, so Hicks simply blocked it out. During the day he never drew the vertical slatted blinds, made of polished sycamore and inspired by those in the ante room at the Soane museum at 13 Lincoln's Inn Fields, London. Nor did he draw the voluminous silk-taffeta curtains at the windows, yet the feeling in the apartment was that of lightness and brightness despite the Vandyke-brown painted walls. "Rich colours absorb light," he observed, "but the ceiling painted white reflects light massively." Glass bulbs (lamps) cast shadows on the ceiling, so Hicks bought ones with built-in reflectors. He further controlled the spread of light with his own home-built baffles – inside every lampshade he fitted angled semicircles, cut from silver or gold card, to let light wash exactly where needed.

Left David Hicks never liked overhead lighting in bedrooms so he installed lots of sockets for table and floor lights as well as freestanding halogen uplighters placed out of sight. Small fibre-optic bedside lights with dual controls are fitted into the bed, with the box that powers them hidden underneath. The Regency light (shown in the foreground) is powered by incandescent candle bulbs (lamps).

Right Table lights and wall uplighters give a soft, diffused light, so Hicks cut semicircular discs to fit inside their shades to direct the light. Downlighters recessed in the ceiling wash the dark-coloured walls with light. A white ceiling bounces the light back from halogen uplighters hidden behind the sofas.

Right In the reading room, access to the Ruskin Library's contents is from computer terminals equipped to locate and order each volume. The entire contents of the library are on CD-Rom, so the desks are wired for laptop users. Light specifications here are carefully controlled, not only to protect the volumes, letters, and paintings, but also to allow screens to be read in dimmer light. Now that information is conveyed in pixels of light glowing on screens, we need less light to read it by.

Left An innovative design solution sheds carefully controlled light on the world's largest collection of the work of John Ruskin (1819–1900), 19th-century English author, artist, social reformer, and conservationist. In the Ruskin Library, architects MacCormac, Jamieson, Prichard allowed light to enter the windowless cylinder through tall, narrow apertures at the entrance and at the far end, where the reading room is found.

Right Nothing is allowed to block the passage of light from reaching where it is needed. The glass and slate floor links the entrance and the reading room in the Ruskin Library. The glass bridges at first-floor level between the galleries were inspired by Sir John Soane, and have the ability to beam light right down into the basement.

Richard MacCormac

In an age of bland, overbright lighting, British architect Richard MacCormac, of the firm MacCormac, Jamieson, Prichard, is inspired by the dramatic effects of Soane. "We have become so used to the ambient energy of electricity that we have lost any sense of what it was like in the pre-electric age," he believes. The starting point for his own lighting schemes is the 18th-century artist of light and space, JMW Turner, and the architect Soane. Their interest in the experiential effects of space and colour relies as much on chiaroscuro as a steady brightness, and MacCormac introduces that play of light and shade into his interiors. Soane's domes inspired MacCormac's vaulted toplit public rooms at St John's, Oxford, where white concrete structures and coloured walls are concealed beneath the upper terraces. The monochromatic toplight that bathes Fitzwilliam Chapel, Cambridge, creates a calm space. Windows make space habitable, which is why MacCormac refers to "inhabited windows" – including the bay windows at Worcester College, Oxford, and those of Trinity and Fitzwilliam colleges, Cambridge.

MacCormac's Ruskin Library at the University of Lancaster, England, is perceived as a type of lantern – a beacon to students – which explains why there are no windows in the cylindrical building. Instead, light beams from the bold, double-height, glazed entrance, which is repeated at the far end. MacCormac's masterly manipulation of light and space means that the building becomes the library for the 21st century, with light levels that are suitable for reading laptop screens – essential, since all of the library's archive material is stored on CD-Roms – even though, at the core of the building, the Treasury Box houses all of the original volumes and manuscripts.

Once admitted through the glazed apertures that are sliced out of the windowless cylinder, light is bounced around the space in a narrow stream by reflective surfaces, which the architect describes as "Byzantine in colour, Gothic in mood, a metaphor for Ruskin's Venice". Bronze doors, clear glass, and slate floors, plastered with a translucent Venetian red, reflect the light. They also confuse the precise depth of focus – another Soanian trick – and the effect within is the dissolution of physical enclosure.

At home with a lighting consultant Sally Storey

As the design director of two lighting companies, John Cullen Lighting and Lighting Design International, specializing in large commercial projects such as shopping centres, museums, and hotels, Sally Storey knows all about bringing her work home at night. In her London house she shows how modern lighting can enhance Victorian architecture. A low-voltage downlighter concealed in a ceiling rose brings more sparkle than glare to the centre, unlike the ubiquitous pendant that once hung there. In old gaslight wall sconces there are now halogen wallwashers.

Sally Storey likes low-voltage halogen because it is cooler than incandescent's warm light, and, when dimmed, it still provides a glow for the evening. Low-voltage light fittings operate on an electrical supply of 12v or less, so they are very energy efficient. With the multimirror dichroic system that lines the cap, the bulb (lamp) passes heat backward rather than projecting it outward. In the John Cullen Lighting collection there are up- and downlighters with different pin sizes, bayonet caps, wattages, and beam angles for spots, as well as a choice of silver or gold reflectors.

Storey's approach to lighting is architectural. First, accent light is produced by directional downlighters, and low-voltage recessed halogen fittings are positioned long before decoration takes place, with their transformers hidden in the ceiling. She never arranges these downlighters in a symmetrical grid. Rather, their location is dictated by what needs to be highlighted. Second is uplight, for which she uses low-voltage halogen, either installed or freestanding, shining into ceiling corners and onto the front of large pieces of furniture. The lights can be hidden behind the furniture or built into it. Third, ambient light is provided by table lights, but never more than four in a room because they look lifeless switched off, or

Far left Prioritize what you need to light. The focal point of the drawing room – the fireplace – is illuminated at night by a pair of wall sconces from McCloud & Co, using low-wattage candlelight bulbs (lamps) and paper shades.

Left At night, wide-angle, low-voltage halogen "Polestar" downlighters from John Cullen Lighting, recessed in the ceiling, bathe the wall in light and boost the illumination from the sconces. A matched pair of downlighters are installed on the far side of the room.

Left Hidden in the drawing room ceiling are seven halogen downlighters, three on opposite sides of the room, and one in the central ceiling rose – its beam narrowed by honeycomb louvres to reduce side glare and the spread of light. All four corners are lit by freestanding low-level halogen uplighters. Table lights help to layer the illumination.

Below At night, lights inside make the outside appear darker as the windows mirror that blackness. To counter this, Storey always lights the space outside with a 12v fitting buried in the ground. She even lights the outside perimeter of conservatories, using only candlelight within the space.

floor lights with flexible swivel heads dotted about the spaces where they are needed.

Storey's brief to the electrician is to put outlets in all corners of main rooms for freestanding halogen uplighters. Hidden by furniture, these little "softeners", as she describes the "Pockets of Light", have three beam widths: narrow for paintings; medium to wash light upward over a tapestry or for under an indoor tree; and wide to be used behind a sofa where the glare can be kept well away from the eyes. These light sources can be moved around, unlike ceiling fixtures, but they need outlets in the right places. Light switches in a Storey scheme are always set lower than normal at about 3ft (90cm), in line with door handles, where they are less obtrusive.

Storey believes that good illumination is all about layering light, using different intensities and sources. It is also about understanding light and shade. Lights controlled together on the same circuit without dimmers produce uniformly bland, and often overlit, interiors. "Three well-placed, low-intensity lights on dimmers are better than one bright light that glares at you and is unflattering," Storey says. So she installs four circuits in the main rooms. She also briefs the electrician to install dimmer circuitry and a preset computerized programme that saves twiddling knobs.

Light levels on the four circuits change automatically between a cool white light of daylight, a glow in early evening, a warmer glow at night when more incandescent lights switch on, and finally a low light for the late evening that alternates shadows with pools of light, like a Caravaggio painting. If the mood is not what she likes, she can override it. "I vary the focus to get an interesting balance, the way that daylight changes," she says. Artificial light must be variable. Without incandescent table lights, colour and texture flatten, so she builds in enough outlets for them. In contrast, most of her rooms have seven built-in, low-voltage, halogen ceiling downlighters, positioned to concentrate light on a windowless wall or some other feature that needs highlighting, such as the middle of the dining table or the centre of the ceiling rose that is a common feature in conversions of older houses.

Lighting is also about creating certain moods, not just the science of throwing enough – never too much – light onto the subject. So it is useful to observe

Left In the library, "Columini" uplighters built into the bases of the pilasters on the bookshelves pick up the detail of the painted architrave. Overhead, a "Polestar" ceiling light with a narrow beam picks out the bindings of the books.

Below Terracotta-coloured library walls and a textured seagrass floor, with colourful rugs underfoot, are lit by four halogen medium-beam and three narrow-beam "Polestar" downlighters in the ceiling. In addition there are three "Porta Romagno" table lights with parchment and cream silk shades.

Left "Pocket of Light" is a useful twisty, bendy 12v floor light that hides its transformer in its bronze or nickel-plated base. Its halogen lighting head swivels and can be focused to give three beam widths – wide, medium, or narrow – depending on what needs to be illuminated.

Below Dining room walls painted ochre glow with indirect light from the narrow beam of a "Mira" inside the glass-fronted obelisks. Above the table, the narrow beam of a "Polestar" balances the candlelight, while a medium-beam "Polestar" lights the picture. A further seven halogen down-lighters are recessed in the ceiling.

Right "Columini" uplighters built into the base of a set of bookshelves highlight the pilasters. A "Wotan" metal reflector from Osram prevents any glare from the narrow beam width of 10 degrees. Even up close, you are not blinded by the light.

how everything, including the kitchen sink, is glamourized with light effects. Just as decorators use accessories to layer texture, Storey layers lights, dimming down the boring bits while accentuating others with focused lighting. You do not actually notice the fittings but the effect is there. Storey believes "lighting is about effect, not source", which is why table lights interest her least in the overall scheme. Her lighting is always flattering to texture and colour and to people, even in the utilitarian parts of the house, such as the kitchen and bathroom. In the bathroom, for example, she backs up good, bright-white "Polestar" halogen lights, reflected on white marble surfaces, with side lights more flattering for the make-up artist. Inside the shower, three sealed "Water Lily" lights give the effect of running water on the tiles.

A kitchen needs to be well illuminated for cooking, yet subtly lit for supper. In her kitchen, Storey avoids the commonplace – that cold white fluorescent tube concealed under units throwing a ghostly pallor on the work top. Instead, she uses halogen to light the fronts of kitchen units and reflect the light from "Polestar" ceiling downlighters. Generally, she avoids an arrangement of downlighters, except in a basement kitchen that has to be lit all the time. Over the sink, scalloped light from "Mira" downlighters emphasizes the linear effect of slatted blinds. Four downlighters above the blinds are hidden in the window reveal.

"Often, architects plan light fixtures at blueprint stage on the drawings in a two-dimensional grid in order to achieve an interesting starburst arrangement on the ceiling. They don't think three-dimensionally about the effects of that light. Place your overhead lights exactly where they are needed – don't worry about a symmetrical grouping," says Storey.

Storey likes the coolness of halogen on her red and ochre walls. Even with the white kitchen units, she likes the sparkle that reflected halogen brings. At night she uses a lot of candles with back-up from table lights with parchment shades that yellow as they age.

Right Brightly illuminated by day and softly and subtly lit in the evening for supper, the kitchen has "Polestar" down-lighters recessed in the ceiling beaming their light onto the stone-tiled floor and white-fronted units, which have a satin finish to bounce back cool, white light.

Far left In addition to the ceiling lights, the kitchen also has a "Mira" dichroic wall light recessed into the surface so that you can never see the actual source. It throws out a strongly scalloped light to emphasize the linear effect of the slatted Venetian blinds. There are another four downlighters set into the window reveal above the blind.

Left The range of fittings from John Cullen Lighting used in Storey's house are shown here.
1 and **2** "Mira" low-voltage recessed lights with different wattages.
3 "Polestar" low-voltage halogen fitting, which can be angled to direct the light where it is required.
4 "Highlight" freestanding, low-voltage halogen uplighter.
5 "Columini" low-voltage halogen uplighter.
6 Preset computer programmer for lights with four settings, ranging from bright daylight to a dim, late-evening setting.

and Major like contemporary fluorescent light. As well as its low-energy requirements and low-heat output – always an important consideration where computers are present – it has the soft, diffused quality of light that Major describes as "great for a drawing office". Eight large ceiling-mounted "Trave" fittings from Zumtobel, with two sets of lights directing light down and one set upward for more general illumination, create both direct and indirect light in a single fitting. Two "Modular Nomad 4" units, each using 12v AR11 metal reflector lamps, focus light on the end walls to highlight the texture of the brick.

Desk lights – "Tolomeo" by Michele De Lucchi for Artemide – focus on the printed page, in this case specifications or plans. The output is a warm, incandescent glow from a swivel head mounted on an adjustable arm. As it swivels it changes its performance. The tension created with a spring and a cable, similar to a bicycle brake wire, hidden in the arms, supports the head. The friction, plus the spring in the tube, means that it works in all positions. The effect changes simply by changing the position of the light. At one angle the light is diffused, at another it is spread across the table. Tungsten-halogen spots in the studio give background lighting, turned down low on dimmer controls at night to a warm, domestic glow if the designers are working late. They like to contrast warm with cool light, as they do with polychromatic effects elsewhere in the office.

Outside the meeting room, surfaces are coloured to illustrate the effects of light on paint, and gel filters are employed to colour the light and intensify it. Two standard linear fluorescent battens, both gelled orange with filters, make the cruciform video display panel on the door appear to float. Approaching the entrance to the designers' studio you can see a furnace-like red light seeping under the door. "We wanted visitors to know this was a lighting design studio," Major explains. The red surfaces are lit by four "OLS Nimbus" floor uplighters, using 12v tungsten-halogen lamps fitted with red dichroic filters, and the orange walls are lit by 12 Concord "Instar" fittings using 12v tungsten-halogen capsule lamps. In the library there are four "Modular Nomad 4" fittings, each one using four 12v AR11 metal reflector lamps, to illuminate the table and bookshelves.

Behind the cruciform door (**far left**) lies the meeting room, which, when seen in daylight and with the lighting switched off (**left**), is all white. Chairs, tables, surfaces, even the blinds are white to reflect as much coloured light as possible. The blue-painted niche is used to demonstrate how a different-coloured reflective surface reacts under changing polychromatic lights. An "Irideon Composer" system provides colour-mix and kinetic light from eight "Irideon AR5" colour-change projectors mounted on the ceiling. One "Modular Multiple 3" is used to light the blue niche. These projectors are programmed for three- and two-second light changes – switching from one to another on a preset programme controlled by an LCD touch screen. In the meeting room, changing light effects are simply focused on chair backs (**above right**) to show a variety of colours against a blue background. For example: an orange glow against a night sky backdrop that replicates sunrise (**left**); the red glow of the sun sinking against the dusky blue light of early evening (**right**); and the wintery silver-gray light of the midnight sun (**far right**).

Right Louvred wooden blinds wrap around Starck's home-office outside Paris, France. "The quality of light can affect your life," Starck believes. The blinds control and filter the daylight as well as warming its appearance a little.

Far Right There are no recessed ceiling downlighters – just the incandescent "Romeo Moon" pendant fitting designed by Philippe Starck to produce a diffused light through an etched-glass envelope around the light source.

At home with a
lighting designer Philippe Starck

Designer and architect Philippe Starck has created everything from chairs and bathtubs to boats and motorbikes, as well as designing a chain of hotels in the United States and England. He lives mostly in France in a house close to Paris, with his partner, Patricia Bailer. Built by Starck in 1984 as a family home, the five-storey building became his main office. Three floors for designers, architects, and company management are sandwiched between a more domestic meeting room on the ground floor and the top floor, which is Starck's own room and office.

He has tailor-made his world to be smooth and harmonious, with no signs of ostentation, just as the Eames chair and wrap-around windows on the horizontal line declare his modernism. Because natural light enters through the windows, flooding the house in daylight, surfaces mostly have a matte finish. At night, he builds atmosphere in his house by multiplying

light sources to an incredible degree. His approach to lighting is more high-touch than high-tech, using low-voltage table lights, tiny Christmas lighting decorations, fairy lights, and, eccentrically, more than 70 china sea shells fitted with lights.

"The average lighting designer's approach to specifications and how many lux per square metre is different to mine. I just try to do something nice and sensitive with light. I don't design machines that make light. I never reveal the technical parts that make obvious the source of the light or the mechanism to turn it on or control it. It is impolite," says Starck.

Looking back to the start of his career, Starck produced two lights that hovered in the air in a pocket of helium, but they never got beyond the prototype stage. One took the form of an ectoplasm sac, containing an incandescent bulb (lamp), betrayed only by its umbilical cord; the other

Left and **far right**
Starck's black and
white portraits and
tribal masks, lit with
traditional incandescent
picture lights, reveal his
quirky interests in
humanity – both bare-
faced and masked. Like
an anthropologist, he
likens his audience to
a tribe. "The mass
market is no longer
relevant. Instead, there
are niche markets
which are global.
Designers just need to
speak to our tribes and
be sharper to please
them. This is the real
revolution of pro-
duction," says Starck.

Above Every half an
hour there are 1500
"Miss Sissy" lights
sold around the world.
Designed by Starck,
and made by Flos of
polycarbonated
luminous material, the
entire object glows
when the light is
switched on – not just
the shade on top of the
base. So the whole form
functions as a light
couched in a familiar
little shape – one that
Starck likes so much
that he quips: "There is
even one in the
refrigerator at home."

Above right The horn
shapes that inspired an
entire range of Starck
products, from door
handles to a kettle for
Alessi, are bathed in
slatted daylight.
"Designers must think
before they act," he
believes, "to produce
something useful that
is entirely new, not
merely to find a new
shape to replace
another perfectly
good one. The only
moral way to design
is to try and make
something like the light
without a lamp or use
the same light fitting to
hold different sources
of light in the head."

was a fluorescent tube wrapped in a cushion. Designing a light, he says, means balancing a system whose extremes are defined by the light source and by devices such as screens or diffusion elements that modulate and control the flow of light. Orientation toward one or another of these extremes implies certain differences. Working on the diffusion elements – the shade or reflector cap – brings light closer to being an object. On the other hand, direct manipulation of the light source places designer activity in the spatial scale. In the former, the light appears as an element of set design; in the latter, a prop. The light by **Man Ray**, which spreads its gentle glow on either side of the bed "is the simplest light ever made and thus the most elegant" Starck believes. He likes different light sources, from daylight to candlelight, firelight, and mirrored light: "If you must work from bed, use the computer as a light source." What will make the 21st century different, he believes, is the millions of people working from home. This is why he designed the "Lazy Working Sofa" for Cassina, which doubles as sofa bed and workstation. "Look at the work of the photographer for ideas. The camera never lies. Make more shadows, here and there," he says. His basic belief that the world does not need designer objects any more is why he introduced different heads for different sources – halogen, compact fluorescent, and incandescent – to fit the same light base in the "Archimoon" series for Flos. You choose the head you need for the task at hand.

Above Ingo Maurer's "Birds Birds Birds" is designed around an aluminium pole that anchors the "flying" incandescent bulbs (lamps). When turned on, the fitting seems to fly with a free spirit and an illuminating grace.

Right "Hearts Attack!" is a special one-off celebratory design, created for a feast-day, consisting of 48 plastic hearts and mirrors, each with its own 12v lighting system.

At work with a lighting designer Ingo Maurer

Nobody has done more to make light electrifying than this century's lighting design genius, Ingo Maurer. Cool reason dictates his choice, and a magician's touch illuminates it. He illustrates this yin and yang of opposites with his two collections – the everyday products and the "one-offs" for performance art shows. Both are flattering to the viewer, using a soft light from many sources. His designs always combine indirect and direct light in the same fitting, thus creating a task light that is also able to produce a warm, intimate atmosphere. At the biannual Euroluce lighting show in Milan, Italy, Maurer exhibits his product collection to enthusiastic crowds, while away from the show in the Spazio Krizia he stages a show of startling effects with Ron Arad, Professor of Furniture at the Royal College of Art, London. At the last Euroluce for the 20th century, staged in April 1998, he launched his idea for the light source for the 21st century –

light-emitting diodes (LEDs) – the same cold, intense light that is reflected in car brake lights. In between these lighting fairs, he likes to concentrate on making special pieces intended for limited-batch production. These allow him to create new installations that are more artistic than the products demanded for his regular collections.

He has always been keen on staging parties. In Cologne, Germany, when the furniture fair takes place, Maurer always draws crowds into the 1320-ft (402-m) Deutzer Bruecke Bridge spanning the Rhine. The entrance is through pipes only 4ft (1.2m) high, where light is excluded. The floor tilts as you walk through darkness and disorienting fog and mist from dry ice machines, and then you enter an undersea world immersed in a deep blue light. Everybody passes by an electrically charged field with spirals of free-swimming fish and skates rattling on the sea bed. Giant fish, attended

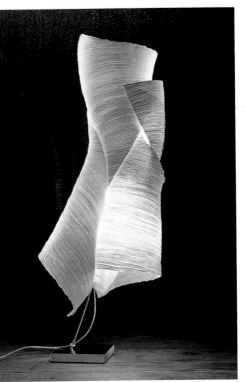

Above "Cuore Aperto" is a hand-crafted aluminium heart, a hybrid of two of his other designs – "Birds Birds Birds" and "One for the Heart" – made for an Italian fiesta.

Left The "MaMo-Nouchies" brings Maurer back to Isamu Noguchi, whose ideas were based on the traditional craft of *akari* paper, and Dagmar Mombach's transformation of an ancient Japanese textile technique.

Right "Fly Candle Fly!", the idea of Georg Baldele, is based on open votive candles and a field of blue and green lights to create a firefly-like path.

Above and **left** Away from the fairs, Maurer's imagination runs free with lighting spectaculars such as this. In the Deutzer Bruecke Bridge venue he staged "Tales of Light" to create the illusion of being on the sea bed, in a soft blue light amid swirls of fog. Dancers dressed in fish costumes, 20ft (6m) long and made of transparent nylon, glide past smaller fish, while radio-controlled cars, disguised as sea-bed creatures, slither on the floor.

Top right Signs, illuminated cones, and controls at the entrance to the Deutzer Bruecke Bridge venue give no hint of what to expect inside, once you are deep under water. Lit by masses of low-level halogen with coloured gels to produce a red glow, like a disco, the light changes to a translucent blue once you descend to the depths under the bridge.

Right A nonexistent light bulb (lamp) in an empty space changes colours at different angles, from red to blue, yellow, and green. The light is, in fact, a hologram of a light bulb (lamp), which consists of 56ft (17m) of film. The actual light source is a disguised halogen lamp.

than as a horizontal one; wallwashers can be used to counterbalance this illusion. Switch off the wallwasher and the beams from the ceiling spots scallop down to give a so-called "cave" effect. Table lights illuminate only to dado height, which is why in high-ceilinged public spaces they are used as task lights on desktops, or as dramatic eye-level accents; Jean Nouvel used "Costanza" lights throughout the glass-fronted Cartier Foundation galleries in Paris in this way. At Nouvel's Institut du Monde Arabe, Paris, only the columns within the building are lit at night, so that the four ceiling lights from Erco ring each pillar and a dark pathway creates a dramatic route through the museum.

Even in stores, where it is necessary to place objects in evenly lit places, areas of homogeneous general light are balanced with less-bright light, and spots, four or five times as bright as the background, bring a focus to bear on close objects. Different shapes are emphasized with different quantities and qualities of light, since uniform illumination flattens out shapes. Photographers know this, which is why they use halogen spots with variable beam widths on curved objects to bring them out of the background. Car showrooms sport a lot of curves and reflective metallic surfaces, so dim background illumination is created with diffused fluorescents, while big-beam halogen spotlights are used to pick up on rounded shapes. Smaller spots focus on specific details.

Incandescent bulbs (lamps) emit heat, which is why you seldom find them used in museums. Halogen spots, which also emit heat and can, therefore, create condensation, would be disastrous inside showcases.

Below left Some 40ft (12m) below the street, the new Southwark station, designed by Richard MacCormac for London's underground network, has daylight playing on its faceted blue-glass wall, designed and silk screened by artist Alex Beleschenko. A computer determined how to wrap 600 glass triangles around an elliptical cone. An above-ground light-well beams light into the station, where it is refracted and reflected by the glass wall.

Below Architect Nico Rensch opted for a low-tech performance, high-industrial approach, with metal pendants and exposed pipes and ducting, for his office conversion of a former factory. Differentiating the areas is as easy as guiding users along a pathway of light.

Bottom Norman Foster Associates commissioned the largest luminaire in the world from Erco for Stansted Airport, England. The fitting itself is hidden.

Right The sculptural quality of the exhibition space at the Guggenheim Museum Bilbao, Spain, designed by architect Frank Gehry, sinuously wraps itself around a sculpture by American Richard Serra on the ground floor. No beams or pillars interrupt the space, and a large central atrium channels light down into the deepest recesses. Glazing around the stairs allows that light to penetrate and walls of light turn into luminous screens.

Left "Darklite Washlite 39" by Edison Price Lighting, used in the Abstract Expressionist galleries at the Museum of Modern Art, New York, can be positioned to provide maximum illumination at eye level.

Objects in showcases are lit with fibre optics carrying light down cabling from a remote source – halogen, halide, fluorescent. Because the heat-emitting fibre-optic light box is far from the object, which can be both precious and fragile, there is no change in the micro climate. The restoration of Leonardo da Vinci's *Last Supper* in Milan, Italy, pioneered a new source of cool-temperature light using plastic film made by the Texan firm 3M. This material channels light evenly along a transparent plastic cylinder of any length without diffusing its intensity. It is also a completely static light, the same at noon as it is at dusk.

Museum light specifications are 75 lux per square metre per work of art. Our retinas will adjust to this low level of light but we feel disorientated if we are plunged suddenly into this environment. To reduce the impact of moving from bright outdoor light into a dim interior, architects create a darkened anteroom first, using it like a frame for a canvas to enhance the experience of moving from one light field to another. Installations are always based on the difference between indoor and outdoor light and the distance from which exhibits will be viewed. At 3ft (1m), the quantity of light needed to view an exhibit clearly differs from that needed when the viewing distance is 6ft (2m). High-pressure metal-halide and tungsten lamps, with their high-luminous quality, throw out a lot of light from just a few sources, which makes them invaluable.

The advent of computers in the 20th century radically changed both the quality and quantity of light in public spaces. Now that the cash register is a computer screen and customer database, and there are promotional videos dotted about fashion stores or car showrooms, light levels have dropped to allow screens to be read. Once uniformly bright interiors are now layered with different levels of light from various sources. In Britain, lux specifications for the workplace dropped in the 1980s from 750 lux per metre to 500. That is a fundamental change to perception in the 21st century. Computers have also made it possible for lighting systems to mimic the change in natural light levels and intensities throughout the day.

Right Richard Meier's Getty Centre, Los Angeles, is a series of pavilions that enclose a courtyard, and are connected by glass-clad walkways. The upper galleries have skylights with louvres to admit and control natural light. Filters remove harmful rays.

Left American lighting designer Claude Engle ringed the base of I M Pei's pyramid at the Louvre, Paris, with Erco uplighters buried in the castle's foundations, angled to beam light onto the steel structure. Sunken halogen lights illuminate the foundations.

Left The steel-ribbed cage of the vaulted ceiling at the National Portrait Gallery, London, designed by architect Piers Gough, brings a beautiful light into the new galleries. Natural daylight is balanced by the focused spots above the glass screens in which the paintings are displayed.

Left The Wellcome Gallery at the Science Museum, London, designed by Richard MacCormac, of the architectural practice MacCormac, Jamieson, Prichard, has a backlit blue wall.

CASE STUDY 1

Architect: Future Systems
Location: North London house built on a narrow site between a pub and a Victorian house
Type of building: Four-storey house, opaque-glass-block façade and transparent glass rollercoaster roof extending at a 40-degree inclination to the other side
Size of property: 2314 sq ft (215 sq m)

Left Outside, wall-mounted spots, by outdoor-lighting experts Bega, shine down on the stainless-steel ramp at the entrance to this opaque-glass-block house. Mounted on the garden walls, 24in (61cm) from the ground, the spots turn the steel ramp into a sinuous ribbon winding around one of the foreground trees.

Right On the other side of the house, transparent glass swoops down to the flagstone terrace at a 40-degree angle to give privacy inside. Lit within by powerful halogen wallwashers, "Floodlight" by Concord, the white, profiled ceiling acts as a reflector.

Exteriors

Built on a narrow, 20-ft (6-m) site, this dramatic house by Future Systems lights up at night like a beacon. The straight-fronted façade of opaque glass blocks swoops up four storeys to roof height, where it changes to clear glass and swoops down at the back to street level. People who live in glass houses always come under scrutiny, but you would need a 40-ft (12-m) cherry picker to see inside this house because of its carefully calculated inclination. Problem-solving is the skill of the architects Jan Kaplicky and Amanda Levete of Future

Systems, who, working with Ove Arup engineers, had to make sure that the house does not overheat and that the inhabitants are protected from bright natural light. Outdoor illumination levels vary from 100,000 lux in bright sunlight to 2000 lux when it is dull. The architects were so skilful that although enormous, electronically operated blinds run the length of the house, the family never needs them except in their bedrooms at night.

Kaplicky points out that the acoustics are perfect in this house. "Water hits the glass ceiling and runs down the sides

and you can't hear it. It's one of the best effects – a film of water in a silent movie."

Since it transmits light, how glass is illuminated indoors is as important as the exterior lighting. The transparent, almost weightless house by day lights up at night, becoming solid as it emits light through its structure. "There is light loss but then again the glass works as a reflector so it balances out fifty-fifty, and it is not a particularly measurable loss," Kaplicky observes. Only two lighting systems are used. Outside, pairs of external halogen spots by Bega are mounted on

the front and back garden walls. Indoors, apart from the owners' designer lights, the only light is from powerful halogen Concord wallwashers with 500w wide beams that play on the ceiling. Painted glossy white, the ceiling reflects the light more than the walls, which are painted silk-matte to bounce illumination back onto the white-tiled floor, like the little ceramic heat-shield tiles pioneered by NASA. For maintenance, wall uplighters are preferable to ceiling downlighters, and they also light the ceiling, which has a special curved profile at the edges.

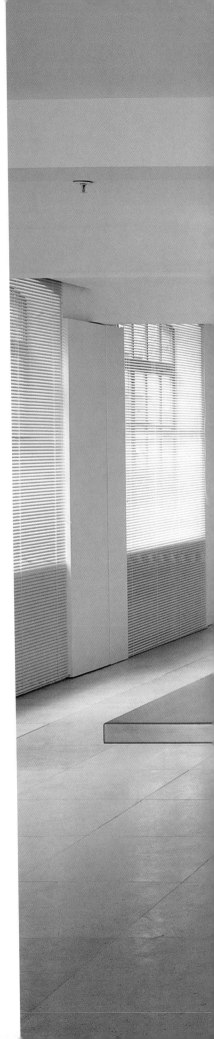

Architect: Mark Guard

Location: Loft apartment in a former factory in Soho, London

Type of building: Originally a tin factory with brick walls and a concrete floor, it later became the location for a famous rock-and-roll club, The Marquee, before being converted to apartments

Size of unit: 980 sq ft (91 sq m)

Right The bathroom, which is formed from the space between the freestanding boxes that divide this loft conversion, is the pivotal point of the apartment. The transparent Priva-lite glass bath screen turns opaque at the flick of a switch. This large piece of glass bisects the custom-built stainless-steel dining table, which becomes the steel tub in the bathroom. A storage wall that runs the length of the open-plan apartment contains everything – including the kitchen sink – and splits into three bays that can be opened as needed. At the far end is the master bedroom, created when the bed folds out of the wall. Even the TV can be pushed into a cupboard.

Open plan

Le Corbusier's prescient description of machines for living in the 1930s becomes virtual reality as the 21st century dawns and factories designed and built in the 19th century to accommodate massive pieces of machinery are converted into apartments. The biggest problem faced by modern architects when they convert industrial spaces into habitable apartments is the need to beam natural light from the distant windows into the building's core. Putting up internal walls blocks daylight, so open plan is the solution. Architect Mark Guard calls this 980-sq-ft

(91-sq-m) unit inside an old warehouse in central London the "transformable apartment" because it is possible to change the configurations of space.

The structure of the apartment is reduced to three freestanding boxes and one 50-ft (15-m) long storage wall.

This long wall opens up to reveal a stainless-steel cooking station, drinks bar, washing-up area, laundry, hi-fi centre, and wardrobe storage. Walls fold out and beds fold down from the freestanding boxes to divide the open-plan space up into a two-bedroom apartment. The bath-

tub is a steel extension of the dining table bisected by a sheet of Priva-lite glass, which turns from clear to opaque at the flick of a switch. This seamless, customized steel fixture and new generation of electrochromic coated glass, whose molecules change from clear to cloudy when an electric current activates them, is a brilliant manipulation of space and light. Standard fluorescent fittings, concealed above the cabinets and the freestanding boxes, uplight the ceiling. To the right of the table is a row of ceiling-recessed, low-voltage halogen downlighters.

CASE STUDY 4

Architect: Claudio Silvestrin

Location: Apartment in a river-front building designed by Sir Norman Foster, in Battersea, London

Type of building: 1980s glazed-front façade – modernist with relatively low ceilings

Size of unit: 3767 sq ft (350 sq m)

The skilful way Claudio Silvestrin has manipulated perspective within this low-ceilinged apartment makes the available space look far larger than it really is. The fully glazed river-front apartment is not overshadowed by any buildings, so too much natural light was the potential problem, especially in the west-facing living room and kitchen. Rather than block the light, the architect's favoured choice is to diffuse natural illumination. A wall of etched glass now screens the kitchen, while etched-glass panels slide across the transparent panes on the glazed façade. In the bathroom, etched-glass discs stop bathers from looking up into a sparkling halogen light source.

Silvestrin's interest in reducing the functionality of objects to make them more expressive "and put a bit of magic and poetry into them" concerns the fitting and the quality of the light. Glass diffusers cut the sparkle from halogen lamps as well as hiding the light source. "Glaring bulbs in downlighters are like going to the dentist. You look up at the ceiling filled with spot-lights looking back at you. I close my eyes."

The apartment has white walls, and to avoid a shiny, reflective surface under-foot and to diffuse the light, the flooring is stone from Serena in Italy.

"The last thing you want to see when you walk into an apartment is the kitchen," Silvestrin observes. Yet leaving the solid wall that originally screened it in place blocked the light from the west-facing glazing and entrance. His solution was a curvaceous, etched-glass wall with a strip of neon to accentuate its shape, which provides a fluid division between the kitchen and entrance.

Silvestrin encouraged the client, Adam Barker-Mill, an artist specializing in light installations, to incorporate his

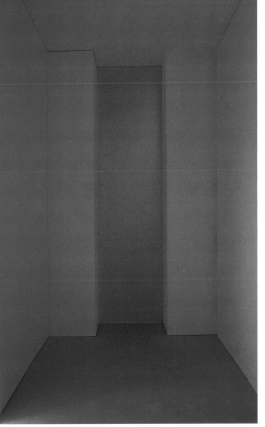

Left A curved wall made from etched glass screens the kitchen without blocking the daylight. A cold-cathode neon strip, just ⅗in (1.5cm) in diameter, has been specially made to follow the curvaceous lines of the free-form wall. Before Claudio Silvestrin's conversion, the visitor to this light-filled space was confronted by a solid wall and darkness. The pear-wood bench and table were designed by Silvestrin; the stone floors are from Serena, Italy.

Above right Rather than curtains, there are sliding screens of etched glass at the river-front windows, and no downlighters.

Below right The intensity of the colour produced by hidden fluorescent tubes makes the light almost tangible in this entrance hall, an effect that distances the visitor from the experience of stepping from the communal areas of a large apartment block into the open-plan apartment. Sealed from the rest of the apartment, this lighting scheme gives the visitor what the architect describes as "totality of experience".

sculptural objects into the architecture. Rather than art, they become structural components built into the columns and walls – permanent rather than temporary.

Entering the apartment is a spatial experience because the architect has created a light chamber that distances it from the dreary, communal areas outside. This small, secret chamber bathed in blue from coloured fluorescents is a transitional room furnished with light. No objects, paintings, flowers – only light made almost tangible by the intensity and density of its colour. "I sought the totality of an experience," Silvestrin reveals. Like a light installation by James Turrell, where you cannot visualize the perimeters of the lit space, he pushes the boundaries by flooding them in light. But his work is not as disturbing as Turrell's. His balanced approach is never dramatic or intimidating. He seeks equilibrium above all else.

CASE STUDY 5

Architect: Eva Jiricna

Location: Knightsbridge, London

Type of building: Brick-and-mortar block built at the turn of the century and turned into apartments

Size of unit: 1615 sq ft (150 sq m)

Living areas

Architect Eva Jiricna opened up a vast apartment by removing a warren of light-obstructing walls and corridors to create a fluid, open living space. North-facing windows fronting the living area look onto Knightsbridge, while the dining area 164ft (50m) away faces south to Hyde Park. Massive doors separating these two areas fold back flat by day allowing daylight to penetrate, but the northerly light at the apartment's core is diffused and sickly. To counter this, Jiricna uses the new generation of energy-efficient fluorescents warmed to an almost natural light. She specified these fluorescents because they compared favourably with examples of tungsten-halogen sources. The fluorescent tubes are hidden behind the original false ceiling, which was installed to conceal load-bearing beams. Jiricna removed the middle section of the ceiling to give extra room height, and illuminated the recess to define the space below.

In a previous refit, load-bearing beams installed to support the floors above introduced these false ceilings, which lowered room heights to a mere 7¼ft (2.2m). Now, where Jiricna has created the lighting recess, the room height is 8½ft (2.6m).

Jiricna uses the illuminated ceiling cornices – the remnants of the original false ceiling – to define seating areas, in much the same way as a floor rug helps to define a conversation area. They also give a good level of background light, to which she adds spots focused on pictures and other objects. "I wanted to avoid those irritating little starburst ceilings you get with halogen spots," Jiricna explains. "Light is one of the qualities that makes life more positive . . . Deprived of light we suffer." The overhead, diffused fluorescent lighting scheme, all the elements of which are controlled by dimmers, is both calming and energy-efficient, while tungsten-halogen spots accent the owners' collection of early 20th-century art.

Surfaces throughout have been chosen for their reflective qualities. Pale stone flooring, with underfloor heating, and polished plastered walls, which revive an ancient Venetian plastering skill, give a silken burnish accented by sparkles from sandblasted glass interspersed with transparent window panes.

Far left Eva Jiricna uses cornices, carved out of the original false ceiling, to hide light sources and create overhead illumination that defines the space for dining and, at the far end of the apartment, the living area. "Wotan" fluorescent tubes from Philips, coated to give almost the same colour as tungsten-halogen light, bathe the silky white-painted ceilings with a good level of ambient illumination. Candles at night bring a warmer glow at table height for a layered lighting effect.

Left While opening up this turn-of-the-century apartment to natural light from the windows, the architect had to retain a reminder of that age – the chimney breast. Shaped like an inverted ziggurat, and highlighted by a pair of ceiling-mounted, low-voltage tungsten-halogen spots from iGuzzini, it becomes a gallery space for a 1920s portrait. An Eileen Gray sofa and Eva Jiricna's own design for interlocking tables are downlit with the same spots from iGuzzini. This north-facing section of the apartment depends on these low-voltage spots and areas of sandblasted glass at the windows to give it warmth and sparkle.

CASE STUDY 6

Architect: Rick Mather
Location: Covent Garden, London
Type of building: Duplex occupying the top two floors of a converted warehouse with an open-plan living and dining space at rooftop level and sleeping and bathing space on the floor below
Size of unit: 2600 sq ft (241.5 sq m)

Inner-city warehouses that were originally constructed for the storage of goods and produce do not always convert easily to become living spaces for people. Problems can be made worse by the often stringent planning restrictions that exist to preserve the street elevations and window configurations of historic or protected buildings. The principal problem architect Rick Mather faced in the loft conversion in this tall, 19th-century warehouse was harnessing sufficient daylight and channelling it inside where it was most needed.

This duplex apartment was arranged with an open-plan living and dining space on the rooftop level, where natural light levels are at their highest, and with the bathing and sleeping areas on the much darker floor below. Rick Mather skilfully toplit both levels of the apartment by installing roof lights in the jagged sawtooth of a roofline and channelling daylight down through glass panels installed in the upper level's maple floor. These Priva-lite toughened glass panels change at the press of a button from transparent to opaque when an electric impulse activates the molecules used to coat their surface. With the glass panels set to

translucent, bathers on the lower floor can see right up to the sky, if they wish.

The three existing pitched roofs were replaced with the same number of generously proportioned skylights, and a fourth was added above the seating area, to create a strip of roof lights that runs the length of the apartment's open-plan living space. Mather describes this spatial manipulation with light: "The shape of the building is irregular, so we tied it together with roof lights that run straight through the middle. They define the space as well as give it a continuity of natural light."

As night falls, the apartment changes. "I like windows which are a bright source of light by day to change at night and dim down as artificial light and candlelight inside take over. Light should never be static," Mather explains. Accordingly, he designed small, square light boxes powered by halogen, and made by Shiu Kay Kan (SKK), and lined them up along the walls where, in a bygone age, the skirting boards (baseboards) would have been positioned. This general low level of light is balanced with fluorescent uplighters recessed along the irregular roofline so that the source is never visible. The roof appears to float above the space.

Right A penthouse area harnesses daylight through roof lights that run the length of this inner-city apartment. Priva-lite glass panels inset in the maple floor channel the light to the lower level, as does the staircase. Motorized blinds filter infrared light and the noonday glare while permitting the entry of high levels of daylight. The kitchen and study are screened behind curvy partition walls, which stop short of the ceiling and so do not block the light. At night, fluorescent "Potential" uplighters on the ceiling are balanced by square skirting board (baseboard) lights, the "3020", designed by Rick Mather, using halogen capsule light sources in a "bizer" box with frosted glass, made by Shiu Kay Kan. "Costanza" fittings, designed by Paolo Rizzatto for Luceplan, balance this diffused mix of warm and cool light with an incandescent glow at different levels.

CASE STUDY 7

Architect: Tadao Ando

Location: The Kidosaki house in Setagaya, Tokyo

Type of building: A residence for three families – the owners and both sets of their parents

Size of property: Site – 6576 sq ft (610.9 sq m); floor area – 5986 sq ft (556.1 sq m)

Tadao Ando is Japan's leading architect, famous for the Church of Light in Osaka and the Daylight Museum in Shiga. He was awarded the Royal Gold Medal for Architecture in 1997 and the Pritzker Prize in 1995. Buildings made of concrete, steel, and glass employ grids based on the dimensions of the *tatami mat* – the traditional modular flooring system in Japanese homes. This is why the floor areas of his buildings are so specific to their sites. The orientation of each of his structures is also site-specific – built to capture every nuance of the changing play of light throughout the day on their "cast-in-place" concrete walls.

This house in Tokyo carries through Ando's uncompromising ideas on un-plastered cast concrete, even in domestic situations. "Natural light from the skylight along the curved walls shows the house breathing along with the movement of the sun or clouds," he explains. Daylight passes into the house from the most oblique sources – not so much wrap-around windows as glass walls and skylights – to be channelled indoors along contoured concrete walls that act like screens, and that cast shadows as the sun moves.

One of the ideas governing the design of Japanese houses is the concept of the outside room, which is why the house wraps itself around the garden. Sunlight on water, walls, glass, and steel forges a link between this house and its garden. At night, the extended concrete frame of the building, holding glazed sections in a grid, frames the trees like a Hokusai print.

Right At night, Ando likes the warm glow of incandescent light, but he changes beam width according to the floor-to-ceiling height of the room. In low-ceilinged rooms he uses recessed ceiling downlighters fitted with 60w beamed incandescent floods with reflectors; in high-ceilinged rooms he changes to 150w types.

Left The house is constructed so that, as the sun moves during the day, natural light falling through the skylight is made visible in the changing planes and volumes of the structure.

Right Ando is famous for his concrete stud walls, which are given a silky, burnished appearance by the natural light that contours them.

CASE STUDY 8

Architect: Julian Powell-Tuck

Location: Riverside One, near Albert Bridge, central London

Type of building: Two apartments combined into one in an eight-storey apartment/office block, designed by Sir Norman Foster, overlooking the Thames

Size of unit: 4306 sq ft (400 sq m)

Right Lights from Albert Bridge create a colourful show within this riverside apartment. Concord "Myriad 111" ceiling-mounted halogen downlighters illuminate the walls and pictures, with halogen reading lights by Christian Liagre on either side of the sofas and a standard light with drum shade by Santa & Cole in Spain.

An elegant apartment in an eight-storey block built on the River Thames in London has superb views of the Albert Bridge across the water, and over the business sector of the city. It was these views that determined the architects' lighting scheme, in particular the final positioning of the downlighters, since the apartment is already luminous with natural light and the glass reflects what is going on outside. At night, reflections of light from the water and the city bathe the apartment with an ever-changing mosaic of light. The lights from the bridge that shine into the water of the Thames glitter iridescently, creating a very special effect, while the lights of cars moving on the opposite bank of the river create a kinetic light show.

When the architects, Julian Powell-Tuck and Angus Shepherd, received the commission to convert two adjacent apartments into this single, vast, glass-fronted home, their main concern was to include that energetic and kinetic light show as part of the scheme, and so they cut the artificial light levels accordingly. "Because the landscape is moving all the time," Julian Powell-Tuck points out "you don't need an internal stimulus." The

small side curtains, which screen the apartment from a neighbouring one, are rarely drawn. Discovering just the right light levels behind such a huge expanse of uncurtained glass was a challenge, mainly because the glass mirrors what is going on all around it.

By dimming the light levels, internal reflections on the glass are avoided. This facility is controlled by a master panel with a console of dimmers that balances the level of illumination inside with the changing light levels outside, a balance that varies depending on the time of day and the weather conditions.

Deeply recessed Concord halogen wallwashers light the walls and the paintings. The depth of the recess of the lights in the ceiling is the key to cutting the level of glare. If the ceiling-mounted light sources in their dichroic reflectors were closer to the ceiling's surface and visible, the windows would sparkle. The architects also set downlighters as far from the window wall as possible and left the white-painted ceiling to reflect the luminous light outside. The surfaces are pale so as not to block the light. Colours range from soft-beige stone, to cappuccino-coloured carpet and honey-blonde woods.

Architect: Bruno Rohrbach, with interior
designer Sandra Rohrbach
Location: Basel, Switzerland
Type of building: Barrel-vaulted building
shared by two families, including
basement with two double garages and
an additional attic storey
Size of unit: 2153 sq ft (200 sq m)

Left and **right** The
"Birds Birds Birds"
chandelier by Ingo
Maurer has goose
feather wings
attached to 24 low-
voltage incandescent
bulbs (lamps), with a
transformer built into
the three slender
ceiling rose rings. The
chandelier invites
users to exercise their
imagination and
creativity. The wires
on which the bulb
(lamp) holders are
mounted, branching
out from a central
aluminium tube, can
be turned, twisted, and
stretched to "fly" in
every direction.

Dining areas

Dedicated dining rooms have been out of fashion for almost a decade as kitchens have gobbled up more and more space, becoming the largest room in the house and doubling as dining areas as well. The recent and welcome return of the dining area heralds a new approach to more flattering lighting schemes.

In this home built by a Swiss couple, one an interior designer the other an architect, the tall, barrel-vaulted ceiling creates a double-height dining room and an airy space for the "Birds Birds Birds" chandelier designed by Ingo Maurer. Soft incandescent light from many directions creates a flattering light for dining as well as a talking point over the dinner table. Open to the sunny southeast, with wall-to-wall glazing like that of a conservatory, the dining area is illuminated by 24 incandescent lightbulbs (lamps) attached to goose-feathered wings that appear to fly in every direction from the central pendant with its built-in transformer.

With such a high, vaulted ceiling, this layered approach to light, with sources on many different levels from the same fitting, is reinforced with candlelight on the table. As the evening progresses, the chandelier gradually dims down and the candlelight takes over.

Big areas of glazing, framed in steel and wood, allow a large volume of light to flood into the house from both sides, giving those living inside a feeling of "being in a garden". The living room faces on to the terrace. Behind it the double-height dining room stretches up to the top floor of the house, and this area, too, benefits from large expanses of light-transmitting glass. A centrally located transparent staircase separates the living area from the dining room, yet, because of its design, it does not block the southern light that penetrates deep into the core of the house, and so makes it possible to look across from one area to the other without obstruction.

The walls in the area above the dining room are made of glass bricks, reinforcing the impression of transparency and lightness. Glass blocks on the top floor of the house emphasize the open-air feeling, which the interior designer owner has encouraged with the simplicity of line and form evident in the furnishing.

CASE STUDY 10

Architect: John Pawson

Location: Architect's own home, west London

Type of building: Victorian terraced house (town house), built over three floors

Size of property: 1750 sq ft (162.5 sq m)

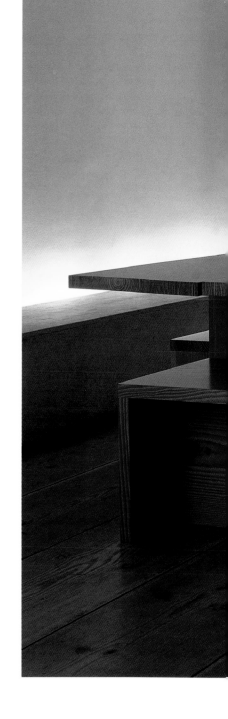

Right It is bold to light the living room with just one continuous tungsten tube uplighter. It runs from one end of the room to the other, on one side only, but it works. John Pawson had the idea to introduce other uplighters, portable ones, if necessary, but it never was. The room is bathed in a sunny light, night and day. Recessed dichroic halogen downlighters are used in the glazed kitchen extension.

Left Pawson believes the light source should always be invisible. A tungsten uplighter in one continuous strip the length of the room is hidden behind a cove in the bench and bathes one wall in the living room with continuous light. This powerful strip of light, warmed up with an opaque-coloured coating on the surface of the tube, is suffused with the glow of the Douglas fir floorboards. The reflected colour lightens as the light travels up the white walls to the white ceiling. As the architect says: "It's a nice diffused light, almost ethereal."

The British architect of Calvin Klein's flagship store on Madison Avenue, New York, John Pawson, converted his own family home in a small Victorian terraced house (town house) in London. It illustrates the maxim that minimalism is an exacting aesthetic. What Pawson looks for is the excitement of an empty space. "It has the capacity to bring architecture alive," he asserts.

Light is an effective tool in enlivening that space, although when there is

absolutely nothing in a room, even the exact position of the light switch requires the most careful consideration if it is not to become a discordant element, disrupting the proportions of the wall on which it is positioned.

Windows are just as important as other features to the flow of space in a room and they are, of course, vital for the control of daylight. The windows, which are at both ends of the house, face east and west. This means that the sun bathes the house in light throughout the day. The kitchen, which backs onto an extension to the original structure, is glazed down the side. The constantly changing patterns of sunlight and shadow throughout the day are the only patterns that Pawson will countenance.

"If the sun is in the right position, you see shadows of leaves on the walls, which creates a beautiful effect," the architect observes. Wooden boards and white walls reflect this.

Every material he chose for his home was carefully considered for the role it plays in reflecting natural light. The floor, for example, is Douglas fir, orange-pink in colour, and the white-painted walls reflect that colour so effectively that they never really appear to be white, but instead are suffused with a warm glow. The ceilings in the house are also painted white so that the walls lighten in their reflected hue nearer the top in marvellous gradations of gray and pink.

Two layers of white blinds installed to provide screening and privacy at the windows also allow the transmission of different levels of light. One set of blinds rolls down from half way up the glazing bar, while the other set travels the full height of the room, from floor to ceiling. The synthetic material from which they are made, known as Primatex, affects the quality of the light. "It is bluey white, like snow – you almost think it's snowing," explains Pawson.

CASE STUDY 11

Architect: Miska Miller working with her husband, industrial designer Ross Lovegrove

Location: West London

Type of building: Brick-and-concrete conversion of a 1958 Richard Seifert leather warehouse into two residential floors and a ground-floor studio for Miller and Lovegrove

Size of property: 4844 sq ft (450 sq m)

Right The windows in the west-facing wall, tilted at 45 degrees, are framed in narrow strips of steel to keep these "light boxes" translucent. A single Philips fluorescent light, set in a Perspex channel in the floor running the length of this 56-ft (17-m) wall, produces a soft, even light.

Far right In the massive open-plan living, cooking, and entertaining space, a pear wood dining table doubles as a light source. Designed by Ross Lovegrove, its legs – the stainless-steel columns at each corner – conceal low-voltage, wide-beam halogen lights and their transformers. Dimmer switches alter colour and brightness. Light reflected from the white ceiling is supplemented by low-voltage halogen spots over the middle of the dining table. The ceiling-mounted speakers (to the right) are part of the sound system.

Tracking the sun with windows and roof lights carefully positioned around the top floor, architect Miska Miller uses daylight to determine the functional areas within her open-plan home. Morning begins with light streaming through two windows in the east-facing kitchen area at the back, where the family eats breakfast. By lunch, the biggest south-facing windows, which overlook a rooftop terrace, bathe the dining table in light. In late afternoon, a glimpse of the low-angled sun through two roof lights is boosted by a fluorescent strip light inset in the floorboards that illuminates a 56-ft (17-m) wall in a soft light. "It makes that west-facing wall appear to float in a soft diffused light," Miller observes.

Furnishing the space with light was the challenge Miller's industrial designer husband, Ross Lovegrove, met with a dining table that doubles as an uplighter. Low-voltage halogen uplighters and transformers hidden in the stainless-steel columns that support the table beam light onto the ceiling, which reflects it back into the room. Dimmer switches in each column allow the light to be tinted and dimmed to suit any occasion. Low-voltage halogen downlighters recessed in the ceiling above the centre of the table make dining atmospheric as well as intimate. And what Lovegrove calls "ambient light" streams down the stairwell skylight from David Hockney's former studio next door.

White walls and ceilings, with a hint of yellow to warm them, and the new oak floor, lightly oiled, are good reflective surfaces – so good that the solid walls and floors appear to float in the luminous space. "It's a brilliant architectural shell," says Miller. "The clarity of light gives you a feeling of happiness". Paper lanterns by Isamu Noguchi and "Costanza" lights with white shades layer the light and fill in the spaces between the uplighters and downlighters.

Modern galvanized-steel window frames were chosen to reinforce the industrial 1950s feel of the original architecture. The floors are wide oak boards salvaged from an earlier conversion, their nail holes plugged with car body filler in fluorescent salmon, purple, and yellow to give a colourful speckle underfoot. There are no paintings on the walls.

Architect: Julian Powell-Tuck
Location: Riverside One, near Albert Bridge, central London
Type of building: Two apartments combined into one in an eight-storey apartment/office block, designed by Sir Norman Foster, overlooking the Thames
Size of unit: 4306 sq ft (400 sq m)

Right On the back wall a glass screen has been installed between the kitchen and hallway, through which lights can be seen. Both the hallway and kitchen were conceived as one space for lighting purposes. Julian Powell-Tuck's own design of halogen reflectors, "Myriad", made by Concord, lights the workstation and three parabolic spotlights illuminate the wall of glass shelves that seem to defy gravity by floating under their heavy load with no visible means of support.

Kitchens

A Bulthaup-designed kitchen located at the back of this city-centre apartment on the Thames in London mirrors in stainless steel the luminous quality of those river-front living and dining spaces seen earlier (see pp. 110–11). Trying to downlight steel work surfaces without producing glare or sparkle is difficult, and it was this challenge that architect Julian Powell-Tuck, with Angus Shepherd, met with his own design of light fittings made by Concord. Powell-Tuck's low-voltage halogen "Myriad" lights give a strong, clear, white light above the workstation,

where the stainless-steel exhaust hood extracts steam and condensation. The exhaust hood, made by Bulthaup, has its own built-in light.

Cantilevered glass shelves, which appear to be weightless and have no visible support springing from the wall, display a collection of heavy flatirons, a display that makes a striking visual pun that the architect chose to highlight with three parabolic spotlights focused on the collection. A fluorescent strip, diffused behind an acid-etched glass plinth at floor level, lights them from below. An alcove

housing the telephone is lit by concealed fluorescent light, and the floor-to-ceiling window opposite (out of shot) is shielded by timber blinds. These lighting effects, and the working wall of flatirons, individualize the mass-produced units and prevent the kitchen from looking as if it came straight from the pages of a catalogue. "Neither modernist or historicist in our attitudes," Powell-Tuck explains, "we see ourselves as gardeners within the existing architectural environment who are unafraid to cut a plant back hard to ensure new and vigorous growth."

CASE STUDY 13

Architect: Peter Marino

Location: Belgravia, London

Type of building: Four-storey, stucco-fronted town house

Size of property: 600 sq ft (55.7 sq m)

Right Task-lighting in this kitchen is concealed underneath the wall-mounted cabinets. Little canopy lights, fitted with translucent diffusers, filter the light and cut the glare. The refraction effect of the filters widens the light beam for a continuous light on the steel surfaces. Low-voltage halogen light in wallwashers and ceiling spots provide a bit of sparkle.

This steel-clad chef's kitchen in Belgravia, London, was first mocked up in New York by the architect Peter Marino so that he could investigate the problems of reflected light on steel work surfaces he would face in a real installation, as well as the effects of lighting on the cork flooring. Multiple ceiling-mounted halogen spots, configured in uneven numbers, give a general, even light throughout the kitchen. The special paint formula used on the ceiling is white, but not too stark. Paint straight out of the can is often too brilliant — especially with so many reflective surfaces in the room. This warmer shade of white gives what Marino calls "a wishy-washy matte surface".

Ceilings reflect light and so Marino cut the amount of light shining back from them onto the cork floor by staining it a dark colour to match the dark brown wood used on the table top.

Ambient light in the kitchen comes from a combination of spots and floods, based on the basic "MR-16" tungsten-halogen lamp, but with different beam widths, backed up with ceiling-mounted spots designed to add a little sparkle to the scheme here and there.

The age-old solution to the problem of providing task-lighting in kitchens by using downlighters recessed in the units has been given a contemporary spin by Marino. He has used translucent glass filters to cut the sparkle on the little "MR" canopy downlighters, which are baffled in the recesses under the wall-mounted cabinets. These translucent glass filters diffuse the light by using reflectors to stop the glare. The glass filters also widen the spread of light in order to provide a continuous and even level of illumination across the surfaces of the steel work tops.

Natural light entering the kitchen also has to be considered. Marino used transparent sliding glass panels in front of the glazed windows to filter the daylight before it enters the room. Set within the recess of the window jamb to enclose the entire glazed surface of the window, these glass panels slide back as needed. Above, 7v tungsten strip lights warm the daylight, which he describes as "typical northern light – white and blue".

Architect: Rick Mather

Location: Covent Garden, London

Type of building: Duplex occupying the top two floors of a converted warehouse with an open-plan living and dining space at rooftop level and sleeping and bathing space on the floor below

Size of unit: 2600 sq ft (241.5 sq m)

Left The next door building is only a short distance away across a narrow, inner-city lane, but bathers cannot be seen. Architect Rick Mather has kept the original windows and covered them with opalescent glass panels, backlit with fluorescent light sandwiched between.

Right Overhead, warm-coloured tungsten spots (only a few, though, because overhead light can cast unflattering shadows) are balanced by the generally diffused light from a backlit window wall. Above the shower, a panel of Priva-lite glass, set in the maple floor on the upper level of the duplex, channels natural daylight down to the bathroom and allows the client to look up at the sky as he takes a shower.

Bathrooms

When a 19th-century warehouse was carved up in the 1980s, some of the resulting apartments were starved of natural light. The task facing architect Rick Mather in the 1990s was to take a saw-toothed, hip-ended roof and beam light – via the upper, penthouse floor of a duplex – into the bathroom and bedrooms on the floor below. An opening in the ceiling – in-filled with Priva-lite glass, which turns from transparent to opaque whenever electricity activates molecules in the coating – became the route for daylight from the rooftop level.

Set into the maple boards of the top floor, they beam that light down to the bathroom. Now it is possible to look up from the shower to see the roof terrace greenery and beyond. This allowed Mather to fulfil the client's desire to see the sky when he showered.

Overlooked by neighbouring buildings in a narrow lane, the windows in the bathroom, vital for the diffusion of light, also needed to provide privacy. Rather than block them up, Mather fixed opalescent glass panels over the existing window panes, with a fluorescent strip light sand-

wiched between the two to create an opaque window above the basins. Now nobody can look in, but daylight levels outside still influence the quality of the light throughout the day.

Warm and cool light are balanced in the bathroom by mixing low-voltage incandescent spots with fluorescent light. Above the glass surround to the basins, tungsten spots producing a warm light are counterbalanced by the fluorescent up-lighter in the glazed window, which, in turn, compensates for the shadows cast by an overhead light. Too many spots look

like a lighting supplier's showroom, the architect believes, just as shiny white reflective surfaces in bathrooms can be unflattering. Only the green glass worktop around the basins reflects light and, below it, a Florentine stone floor, which houses underfloor central heating, is both warm to stand on and warm to look at.

Originally built as a warehouse, the building's conversion to domestic use required some structural changes. For example, the property developer's enthusiasm for a jacuzzi meant that the joists had to be reinforced to support tons of water.

CASE STUDY 15

Architect: John Pawson

Location: Architect's own home, west London

Type of building: Victorian terraced house (town house), built over three floors

Size of property: 1750 sq ft (162.5 sq m)

Below In John Pawson's own family house, the bathroom reveals that minimalism can be sensuous. An outside light shining through a white window blind at night creates a light-box effect, proportioned and positioned carefully in juxtaposition with the Spanish limestone bathtub.

Right Light sources are never revealed, which is why halogen downlighters recessed deep within the ceiling – there are two in the bathroom – have their silver rims removed to make them unobtrusive. John Pawson does not like to puncture ceilings with downlighters, but he admits that in the bathroom they are useful. His own design for the mirror – half etched glass, half mirror – has a hidden compact fluorescent built into it to create a soft, diffused, and flattering light.

A bathroom bathed in light in the architect John Pawson's self-designed home proves that minimalism is anything but sterile. The bathtub, basin, and floor, all made of the same material – Spanish limestone – are sandy coloured, quite warm in appearance, and not particularly reflective. This effect is not like traditional white porcelain sanitary ware, which tends to sparkle in direct light.

The walls of the bathroom have had sand mixed into the cement to match the

texture of the limestone features. The room's ceiling is painted white.

Designed on the lines of a Japanese tub – in other words, "just a box" – the bathtub has been generously proportioned to allow the bather to float when it is full of water. A filter positioned low down allows the bathtub to brim over, and the overflowing water disappears through a gap in the flagstone floor.

At the large modern window, a blind does not quite cover the glass. This is intentional and allows the floating bather to enjoy views of the magnolia outside. Pawson supplemented the natural light entering via the window by positioning a light outside, directed to shine through the white blind. At night, the window looks more like a light box diffusing the white light.

A single halogen downlighter, without a rim, recessed in the ceiling above the bathtub reflects the moving patterns of water on the ceiling. Only one other downlighter, also rimless, has been installed in the room, to allow people to see themselves in the bathroom mirror: "I do not like downlighters much," the architect admits. "Occasionally they are useful, but I hate the way they break up the ceiling. Karl Lagerfeld [the French fashion designer] calls them 'light showers'."

Pawson designed a square light mirror, which is half mirror and half etched glass, to diffuse light evenly over its surface. Standard light mirrors are usually vertical, with an overhead light beaming down a half-moon of illumination at the top while the bottom half is often left in shadow. Pawson's alternative offers the viewer a flattering light, a very soft illumination coming from a low-heat, long-life compact fluorescent tube. "I positioned one fluorescent on the mirror and one directly on the line of a second window so that I can shave in daylight or artificial light and have the light source coming from the same direction."

CASE STUDY 18

Architect: Eva Jiricna

Location: Conservation area in Highgate, north London

Type of building: Brick-and-glass house with a cantilevered roof, built in 1957 by Ove Arup and converted in 1994 for new owners

Size of property: 21,528 sq ft (2000 sq m)

Below The movements of the sun are charted as light falls across the bedroom through panels of toughened, sand-blasted glass on three sides of the room. Even the balustrades on the balcony outside are toughened glass so that nothing interrupts the planes of light falling in visible slabs on the floor. When the sun is low, before any artificial light is switched on, the floor and ceiling are like two planes of light trying to meet each other.

Right By night, the geometry of the bedroom changes as the ceiling lights up and channels in the floor, illuminated by fluorescent strips, make the floor and ceiling more distinct. The bed, with built-in low-voltage lights from Artemide on either side and an electronically retractable television at its base, is like an illuminated room within a room. Air conditioning is built into the canopy, a contemporary version of the four poster, but there is no overhead light in this area to dazzle the inhabitants.

Engineer Ove Arup, whose name is synonymous with the Sydney Opera House and other major projects around the world, built this spacious house for his family in 1957. He used technology beyond the 1950s in this family home, incorporating such advanced elements as underfloor heating on the top floor that radiates heat down through the ceiling and onto the ground floor. More than 40 years old, this Highgate house was beginning to show its age when the architect Eva Jiricna was asked to modernize it for the new owners.

Eva Jiricna decided to use the existing structural beams that supported the floor as channels for fluorescent strip lighting. Interrupting these ribbons of light are the columns that support the butterfly roof. Lighting concealed in the soffits washes over the peripheral ceiling sections, defining them and preventing them from becoming lost in shadow.

Backlighting minimizes reflections from glass, so surfaces are illuminated by hidden light sources, rather than being accented with spotlights. And in order to provide flexibility and control over the performance of the lights, each one can be dimmed on its own circuit. This means that in the morning, for example, there is one set of artificial light levels, while at night there is another.

Daylight in the bedroom pictured above enters through three floor-to-ceiling glazed walls. Even the balustrades on the balcony beneath the cantilever roof outside are large, toughened, sand-blasted glass panels that do not interrupt the view or block any of the light that moves around the room during the day. To achieve this result, Jiricna studied the effects of the daylight by setting up a camera in this bedroom to record the movements of the sun.

"When the sun is low, the ceiling and the floor become two planes of light which give the impression of trying to meet each other," the architect observed. This manipulation of space is enhanced with artificial light when the large windows are screened with mesh glass-fibre and black-out blinds, both electronically operated. By day, with the blinds up, the floor and windows are important, and at night, when the illuminated ceiling defines the space, the room takes on a different geometric form.

The bed, which Eva Jiricna designed, houses not only the occupants but also a state-of-the-art retractable television in the footboard. Small, low-voltage reading lights from Artemide on either side of the pillar supports are ideal for bedside reading. At night the lights of London are spread out at the occupants' feet and it is possible to maintain privacy simply by using the built-in controls to dim the lights illuminating the bed, thus creating a room within a room.

CASE STUDY 19

Architect: Michael Graves

Location: Architect's own home in Princeton, New Jersey

Type of building: Converted Tuscan-style warehouse, built in 1926 by Italian stonemasons working at Princeton University. Originally divided into cubicles for students' furniture storage

Size of property: 6000 sq ft (558 sq m)

Left Overhead lighting causes shadows to form on work surfaces, so Michael Graves uses a flexible arm light by Europa to focus on the desk. Good task-lighting, well balanced with general illumination from wall lights, avoids the problem of eyestrain.

Right The library, with its 17th-century tondo engraving, is bathed in natural light from the skylight that spans the length of the room. Ceiling downlighters, "Torino", are Michael Graves' own design for Louis Baldinger and Sons. The "Torino" has an octagonal, white opalescent glass shade fixed to the ceiling with a satin chrome frame.

Studies

The lighting for "post-Guttenberg" (as the Canadian writer Marshal McLuhan defined the electronic age) man with a study for computerized on-line screen work must be lit completely differently from the period-piece library. The New Age of Enlightenment, dawning as electronic circuitry quietly clicks in to give us knowledge, is illustrated in this house. It has a neo-classical library and a workplace studio adjoining Michael Graves' bedroom on the top floor.

The Warehouse – as the architect Michael Graves refers to his L-shaped home next to Princeton University, where he lectures – was built in 1926, and was given neo-classical and post-modernist flourishes by its owner.

First is the library, which is long and narrow – a shape that is emphasized by handsomely proportioned shelving and the tall windows that formally cleft the walls and look out onto Italianate gardens. By day the library is lit with an overhead transparent canopy, and by night with the octagonal glass-framed "Torino" ceiling light, which beams light down a long funnel onto neatly arranged bookshelves that are stuffed with thousands of volumes on the fine and decorative arts and glossy auction catalogues. Lights are arranged on the ceiling in orderly rows facing each other so that they cast the type of even light necessary in a room such as this.

Next, Graves' small study on the top floor adjoining his bedroom is furnished in blonde-wood shelves. The low ceiling in the attic was a challenge. In order to disguise as well as to make the space seem taller, Graves cleverly dropped a deep box around the perimeter of the attic and threaded all the wiring and plumbing through it. Regularly spaced vertical slits set high on the walls vent both heat and air conditioning and make an ornamental feature as well.

In 1983 Graves needed special lighting fixtures for the Humana Building in Louisville, Kentucky, and approached the manufacturers Louis Baldinger and Sons, who custom made them. This collaboration resulted in the Graves' collection of pendant fixtures, ceiling fixtures, and wall sconces, which are made under the exacting standards that Louis Baldinger established more than a century ago.

1 "Varigola" by Barovier & Toso for Barovier & Toso (1996) Table light with glass stem and base and polycarbonate shade; incandescent, 100w; standard voltage.

2 "Linda T2" by Roberto Pamio for Leucos (1994) Table light in blown, satinized glass, nickel-plated metal; incandescent, 60w; standard voltage.

3 "Soda" by Jonathan Daifuku for Blauet (1997) Ceiling light with blown glass diffuser; incandescent, 100w, halogen or fluorescent, 150w; standard voltage.

The glass tradition

Glass is the perfect container for electric light, which is why light sources – incandescent, fluorescent, halogen, and halide – come packaged within glass. To diffuse or reflect that light is as simple as adding a glass shade, as Louis Comfort Tiffany discovered in the late 19th century when he switched the White House on to electric light covered with Tiffany shades. Early 20th-century modernism moved away from that type of romantic styling, as architects Eileen Gray and Josef Hoffmann concealed light sources in opalescent cubes or globes. Now, anti-glare shutters, UV absorbers, anti-dazzle screens, coloured filters, and beam reflectors control light output. So Murano-glass chandeliers of the 17th century, which made candlelight sparkle, are reinvented to control electric light with dimmer switches and hidden controls, while Philippe Starck's "OA" lamp for Flos hides coloured light sources inside a glass vase that lights when its stalk is tapped.

4 "Selis" by Renato Toso and Noti Massari for Leucos (1994) Wall light in blown glass; incandescent, 150w; standard voltage.

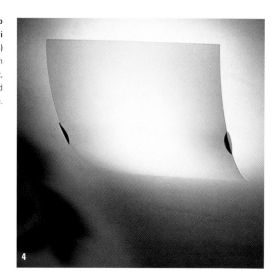

6

7

5 "Lightweight 2" by
Tom Dixon for
Foscarini (1995)
Ceiling light in
lacquered metal with
blown-glass diffuser;
halogen, 8 x 20w per
tier; 12v. Transformer.

6 "KAP" by Barovier
& Toso for Barovier &
Toso (1998)
Floor light in chromed
steel with glass
diffuser; halogen,
250w; standard
voltage. Dimmer.

7 "OCI" by Rodolfo
Dordoni for Flos (1996)
Table light in steel and
pierced glass;
halogen, 75w; 12v.
Transformer.

8 "OA" by Philippe
Starck for Flos (1996)
Table light in Murano
glass with dichroic
reflector; halogen,
35w; 12v. Transformer.
Touch dimmer;
directional flower
head.

9 "Corda" by John
Hutton for Donghia
(1995)
Table light bases in
blown glass;
incandescent, 75w;
standard voltage.

8

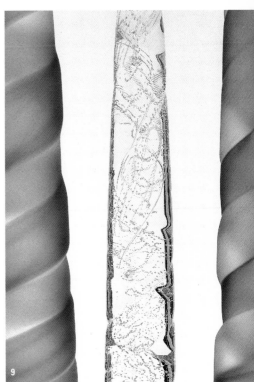

9

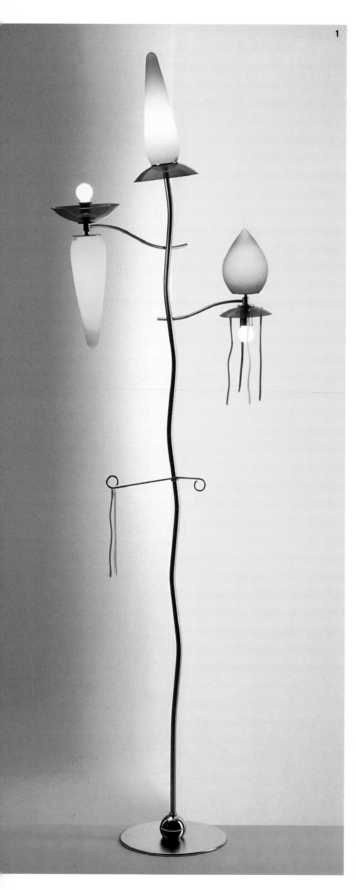

1

2

3

**1 "Giocasta Terra" by
Andrea Anastasio for
Artemide (1997)**
Floor light in chrome-
plated steel, coloured
blown-glass diffusers;
incandescent, 2 x 60w,
3 x 40w; standard
voltage.

**2 "DIY Chandelier" by
Sophie Chandler for
Sophie Chandler
(1995)**
Ceiling light in
Perspex, chromed
metal and blue
recycled glass bottles;
incandescent, 60w
globe; standard
voltage.

**3 "Esa" by Lievore
Asociados for
Foscarini Murano
(1996)**
Table light in blown
glass and chromed
metal; incandescent,
100w; standard
voltage.

4 "ABACO 923.00" by Monica Guggisberg and Philip Baldwin for Venini
Floor light; glass spheres; incandescent, 5 x 40w; standard voltage.

4

5 "Giò Ponti Chandelier" by Giò Ponti for Venini (1946)
Chandelier in multicoloured blown glass; incandescent; 12 x 60w; standard voltage.

6 "Aircan" by Christophe Pillet for Mazzega AV (1998)
Table light/night light in cased blown glass and chromed metal; compact fluorescent, 60w; standard voltage.

5

6

7 "Formosa" by Ettore Sottsass for Venini (1989)
Chandelier; multi-coloured glass and gilded brass; incandescent; 100w; standard voltage. Diffuser.

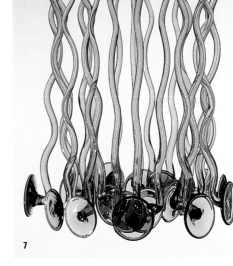

7

8 "Hola" by Roberto and Ludovico Palumbo for Foscarini Murano (1997)
Wall light, chromed or lacquered metal, glass diffuser (white, yellow, or orange); halogen, 150w; standard voltage.

9 "Luxy 269" by Hans Peter Weidmann for O Luce (1998)
Table light in glazed anodized metal with opaline opaque Murano-glass diffuser; incandescent, 2 x 40w, 150w; standard voltage. Touch dimmer.

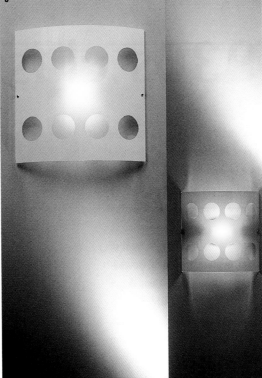

8

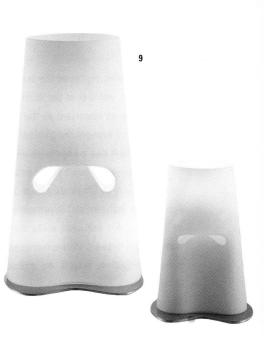

9

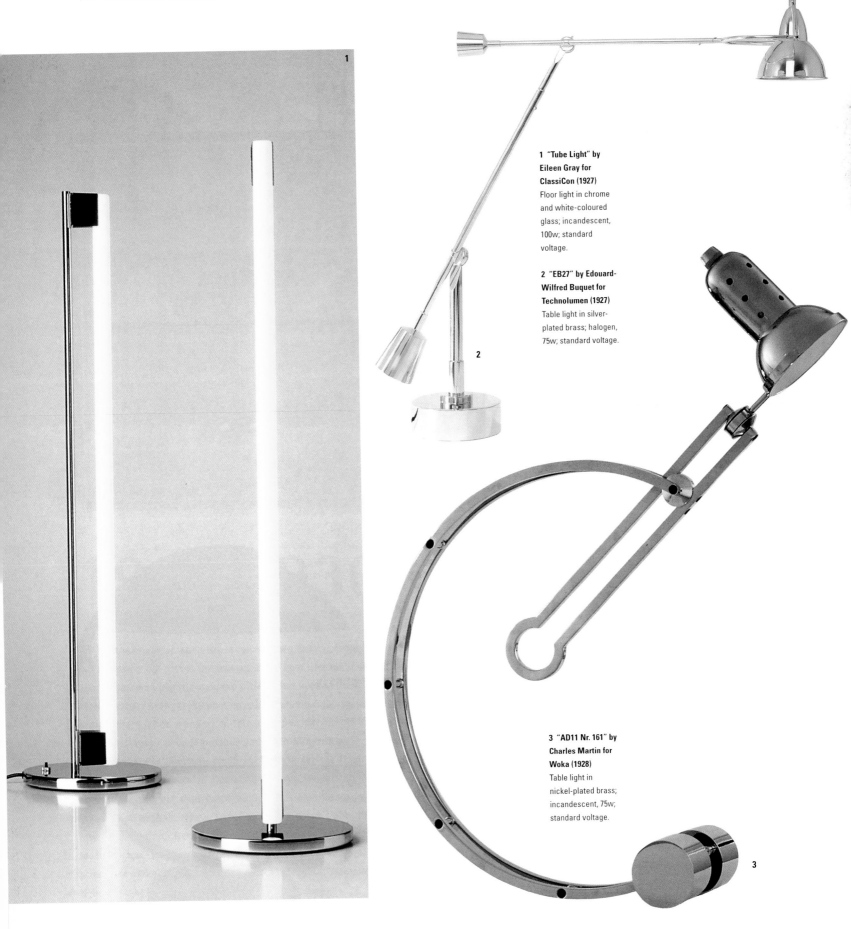

1 "Tube Light" by Eileen Gray for ClassiCon (1927) Floor light in chrome and white-coloured glass; incandescent, 100w; standard voltage.

2 "EB27" by Edouard-Wilfred Buquet for Technolumen (1927) Table light in silver-plated brass; halogen, 75w; standard voltage.

3 "AD11 Nr. 161" by Charles Martin for Woka (1928) Table light in nickel-plated brass; incandescent, 75w; standard voltage.

4 "Art 598" by unknown designer for Alivar (c 1928)
Table light in brass and blown opaline glass; incandescent, 60w; standard voltage.

5 "PH6 1/2-6" by Poul Henningsen, Ebbe Christensen, and Sophus Frandsen for Louis Poulsen (1929/1980)
Ceiling light in aluminium; incandescent, 500w; standard voltage.

7 "L-1" by Jac Jacobson for Luxo Italiana (1937)
Table light in aluminium and steel; incandescent, 50w; standard voltage.

8 "Luminator" by Pietro Chiesa for Fontana Arte (1933)
Floor light in nickel-plated brass; halogen, 500w; standard voltage.
A torch's tallow-dipped wooden shaft is transformed into a sculptural electric uplighter in nickel-plated or brown-varnished brass.

6 "The Anglepoise" by George Carwardine for Herbert Terry/ Anglepoise (1933)
Table light in black aluminium and steel; incandescent, 60w; standard voltage.
Auto engineer Carwardine based "The Anglepoise" on the principles of the human arm to provide spring-tensioned positioning.

1 "Bis 2280" by Fontana Arte for Fontana Arte (1950)
Table light in brass and black and white glass; incandescent, 4 x 40w; standard voltage.

2 "Luminator" by Achille Castiglioni for Flos (1954)
Floor light in stove-enamelled steel; incandescent, 300w; standard voltage.

3 "PH Snowball" by Poul Henningsen for Louis Poulsen (1957)
Ceiling light in white aluminium; incandescent, 300w; standard voltage.

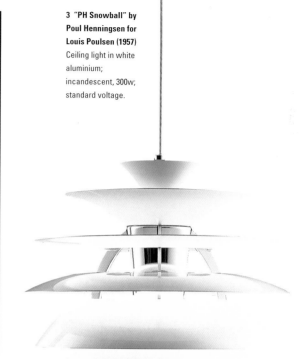

Mid-century classics

The aircraft industry, which had a great influence on designs in the years following World War II, gave lift-off to light fittings that became lighter not only in weight but also in the strength of their illumination. Lowered wattages were used to power lights diffused in nylon, stretch vinyl, fibreglass, and reinforced plastic. Isamu Noguchi gave light a sculptural form through his use of paper lanterns. Scandinavia dominated designs with friendly lamps in organic shapes named "Artichoke" and "Snowflake". Science fiction inspired a series of floor lights that hid their light sources in discs or caps perched on top of tensile-steel stems. The Castiglioni brothers, Achille and Pier Giacomo, brought the spotlight to the Italian lighting industry with "Toio", which was improbably cobbled together from a telescopic fishing rod and car headlight to give the world its first high-tech uplighter. By the 1960s, Italian manufacturers Flos and Artemide were in production with playful pop-art forms.

4 "Alfa" by Sergio Mazza for Artemide (1960)
Table light in silvered metal with glass shade; incandescent, 60w; standard voltage.

5 "1853" by Max Ingrand for Fontana Arte (1954)
Table light in white glass. Two 60w tungsten bulbs (lamps) light the base as well as the shade; standard voltage.

**6 "PH Artichoke" by
Poul Henningsen for
Louis Poulsen (1958)**
Ceiling light in
aluminium;
incandescent, 500w,
standard voltage.
*The inside surface of
the leaves are coated
in a pale rose finish to
warm the light. For
Henningsen
functionalism did not
have to be cold.*

**7 "Splügen Bräu" by
Achille Castiglioni for
Flos (1961)**
Ceiling light in
polished aluminium;
incandescent, 100w;
standard voltage.

**8 "Taccia" by Achille
Castiglioni for Flos
(1962)**
Table light in
enamelled metal and
glass; incandescent,
100w; standard
voltage. Reflective
shield.

**9 "Arco" by Achille
Castiglioni and Pier
Giacomo Castiglioni
for Flos (1962)**
Floor light in stainless
steel and polished
aluminium with white
marble base;
incandescent, 100w;
standard voltage.
Reflector.
*The anchoring marble
base of "Arco" weighs
more than 100lb (45kg).*

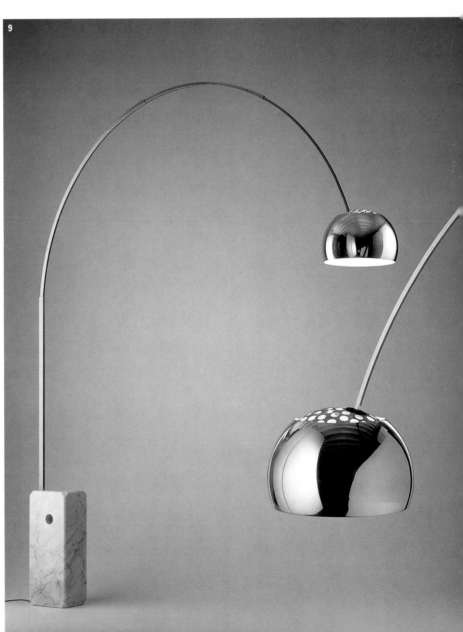

2 "Nesso" by Gruppo Architetti Urbanisti Citti Nuova for Artemide (1965)
Table light in injection-moulded ABS with shade in cellulose acetate; incandescent, 4 x 25w; standard voltage.

1 "Toio" by Achille and Pier Giacomo Castiglioni for Flos (1962)
Telescopic floor light in enamelled steel with brass stem; incandescent, 300w; standard voltage. Height adjustable.

3 "PH 4-3" by Poul Henningsen for Louis Poulsen (1966)
Table light in blown opal glass and chrome; incandescent, 150w; standard voltage.

4 "Wegner" by Hans Wegner for Pandul (1962)
Ceiling light in stainless steel and aluminium; incandescent, 100w; standard voltage. Height adjustable, with separate adjustable shade.

5 "Pirrellone" by Giò Ponti for Fontana Arte (1967)
Floor light in white satinized nickel brass; halogen, 10 x 50w, 24v; 300w, standard voltage. Diffuser.
As Professor of Architecture at the Milan Polytechnic in Italy, Ponti helped established the Milan Triannales and the Compasso d'Oro awards and influenced generations of Italian designers.

6 "Eclisse" by Vico Magistretti for Artemide (1967)
Table light in white-painted metal; incandescent, 25w; standard voltage. Revolving reflector.

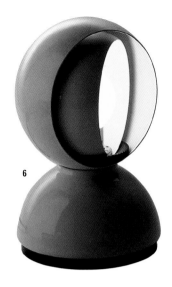

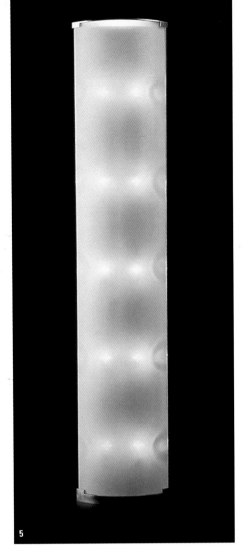

7 "Chimera" by Vico Magistretti for Artemide (1969)
Floor light in white opaline methacrylate; includes tungsten-incandescent linestra tubes, 3 x 120w; standard voltage.

8 "Akari Light Sculpture 1N" by Isamu Noguchi (1967)

and

9 "Akari Light Sculpture UF4-33N" by Isamu Noguchi (1968)
Floor lights in unbleached paper on metal frames; incandescent, 40w; standard voltage.

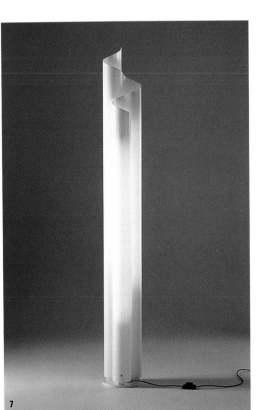

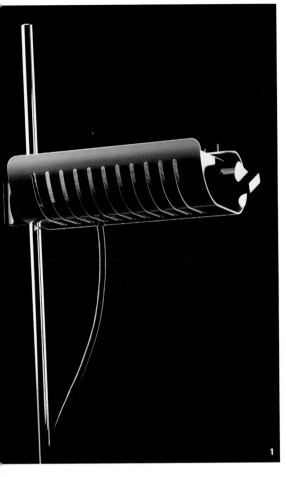

1 "Colombo 626" by Joe Colombo for O Luce (1970)
Floor light in plated steel; halogen, 300w; standard voltage. Diffuser.

2 "Mezzachimera" by Vico Magistretti for Artemide (1970)
Table light in opaline methacrylate; incandescent, 3 x 50w; standard voltage. Diffuser.

3 "Quarto" by Tobia Scarpa for Flos (1973)
Wall light in enamelled metal and clear polycarbonate; incandescent, 75w; standard voltage. Diffuser and reflector adjustable.

4 "Grasl Vulgaris" by Jan Roth for Ingo Maurer (1973)
Table light in aluminium; incandescent, 40w; standard voltage.

1970s classics

5 "Calder" by Enric Franch for Metalarte (1976)
Table light in black aluminium; halogen, 55w; 12v. Reflector.

Halogen, along with other new materials – such as polycarbonates and synthetic resins – brought about a lighting revolution. At last, the miniaturized source with a bright light freed architects to install fittings at the construction stage, rather than add table or floor lights as decorative objects afterward. For the first time the fitting became less important than the light source. The first table light to harness the miniaturized halogen lamp, the "Tizio" by Richard Sapper, was designed in 1972. This fitting holds both positive and negative terminals on the counterbalances on either end of its scaffold-like arms, and the near-weightless head supporting the lamp clips in between to earth them. The decade that began with matte-black minimalism moved on toward a more organic expressiveness, with light fittings in wicker and rattan, wood and paper, and ended with the first signs of a post-modern trend with marble plinths and pedestals supporting light fittings.

6 "Parentesi" by Achille Castiglioni and Pio Manzù for Flos (1970)
Floor light in stainless steel; incandescent spot, 150w; standard voltage. Height and all-round adjustment.

7 "Tip Top" by Jorgen Gamelgaard for Pandul (1971)
Ceiling light in spun aluminium; incandescent, 75w; standard voltage. Height adjustable.

8 "Tizio" by Richard Sapper for Artemide (1972)
Table light in metal and thermoplastic resin; halogen, 50w; 12v. Revolving arm and head.
The first wire-less light houses the weightless halogen source in a tiny head with a sliver reflector. Counterweights and joints make the fitting fully adjustable.

9 "Papillona" by Tobia Scarpa for Flos (1975)
Floor light in metallized prismatic glass and enamelled aluminium; halogen, 300w; standard voltage. Reflector, diffuser; with or without dimmer.
Murano-trained Scarpa diffused the strong beam produced by most uplighters through a translucent glass shield to provide indirect uplighting.

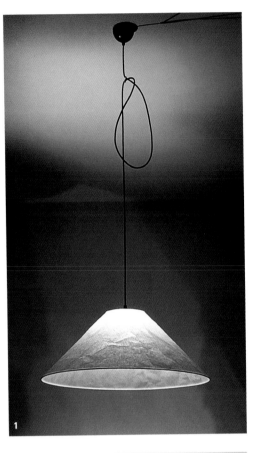

1 "Knitterling" by Ingo Maurer for Ingo Maurer (1978)
Ceiling light in Japanese paper; incandescent, 100w; standard voltage. *Maurer ties a knot in the pendant cord to highlight his twist on the classic decorator's paper shade.*

2 "Frisbi" by Achille Castiglioni for Flos (1978)
Pendant light in opal metacrylate, steel wires and chromed metal; incandescent, 150w; standard voltage. Diffuser.

3 "Megaron Terra" by Gianfranco Frattini for Artemide (1979)
Floor light in white or black aluminium and thermoplastic resin; halogen, 300w; standard voltage. Dimmer.

4 "Bibip" by Achille Castiglioni for Flos (1976)
Floor light in aluminium, ceramic and enamelled metal; halogen, 250w; standard voltage. Rotating reflector. *The ring can be used to alter the direction of the light.*

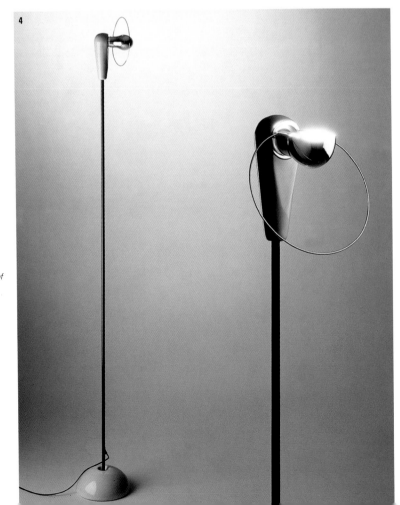

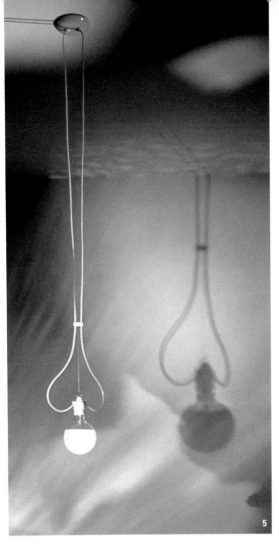

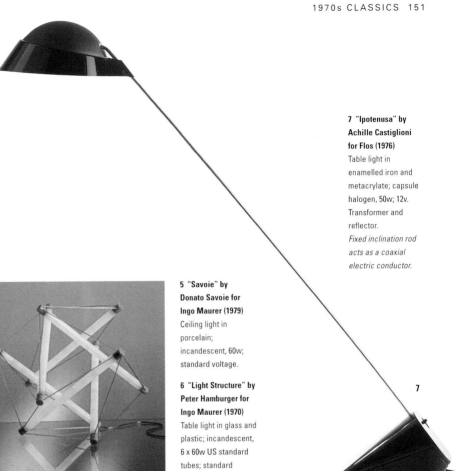

7 "Ipotenusa" by Achille Castiglioni for Flos (1976)
Table light in enamelled iron and metacrylate; capsule halogen, 50w; 12v. Transformer and reflector.
Fixed inclination rod acts as a coaxial electric conductor.

5 "Savoie" by Donato Savoie for Ingo Maurer (1979)
Ceiling light in porcelain; incandescent, 60w; standard voltage.

6 "Light Structure" by Peter Hamburger for Ingo Maurer (1970)
Table light in glass and plastic; incandescent, 6 x 60w US standard tubes; standard voltage. Dimmer.

7

8 "Atollo" by Vico Magistretti for O Luce (1977)
Table light in lacquered aluminium, opaline Perspex or Murano glass; incandescent; 2 x 100w, 2 x 25w for base; standard voltage. Dimmer.

9 "Jill" by Perry King & Santiago Miranda for Flos (1978)
Floor light in glass and iron; halogen, 300w; standard voltage. Dimmer.
A dish-shaped glass shade in a colour of your choice filters indirect light into the room.

8

1 "Tahiti" by Ettore Sottsass for Memphis (1981)
Table light in plastic and metal; halogen, 50w; 12v. Transformer. *The founder of Memphis combined patterned laminates and a zoomorphic form to give a humorous twist on the all-too-serious desk light.*

2 "Aurora" by P A King and S Miranda for Flos (1983)
Ceiling light in glass, Plexiglas and aluminium; dichroic bulbs, 3 x 50w; standard voltage.

1980s classics

Conspicuous for its unbridled consumption, the decade that pioneered "shop until you drop" began when Ettore Sottsass launched the Memphis movement in 1981. An experimental laboratory for ideas, rather than a mass-production unit, Memphis in Italy heralded a playful, colourful approach in which adornment made a comeback, but in a way that was always an intrinsic part of the material rather than being purely decorative. Forms revealed function more amusingly than early modernists envisaged. Light in motion was unleashed by Ingo Maurer with halogen lights flying on cables strung across the room in "YaYaHo". No plugs, switches, or wires are needed, since different lighting elements with dimmer circuits and touch-sensitive controls fly across cables on voltages as low as that for a toy train set. By the end of the 1980s, light was emotive and expressive as designers sought inspiration in magic and fairytales; Philippe Starck's "Arà", for example, is shaped like an elf's cap.

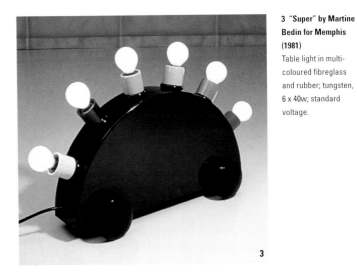

3 "Super" by Martine Bedin for Memphis (1981)
Table light in multi-coloured fibreglass and rubber; tungsten, 6 x 40w; standard voltage.

6 "Eclipse" spotlight range by Mario Bellini for Erco (early 1980s)
Spotlights in anodized aluminium and plastic; low-voltage tungsten-halogen, 100w; 12v. Track fitting; electronic transformer-sensor-dimmer; colour and UV filters.
A modular system with interchangeable lighting heads allows selection of different beam characteristics best suited to a particular application.

4 "Swing VIP" by Jorgen Gamelgaard for Pandul (1983)
Table light in brushed stainless steel, aluminium, and cast steel; incandescent, 100w; standard voltage. Shade and arm pivot 360°.

5 "Kandido Tavolo" by Ferdinand Porsche for Lucitalia (1982)
Table light in black aluminium and technopolymers; halogen, 50w; 12v.

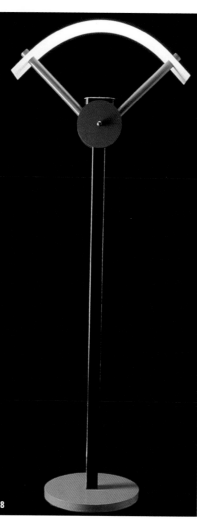

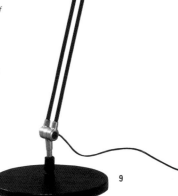

7 "Gibigiana" by Achille Castiglioni for Flos (1981)
Table light with support in aluminium and enamelled metal; halogen, 20w; 12v. Dimmer and metal reflector.

8 "Grand" by Michele de Lucchi for Memphis (1983)
Floor light in plastic and metal; halogen, 300w; standard voltage.
Memphis member de Lucchi believes the task of the designer is not merely to shape the fitting but to consider the quality of light shed, which should always be diffused.

9 "Snodo" by Hannes Wettstein for Belux (1980)
Floor light in enamelled aluminium; incandescent, 100w; standard voltage.

7

8

9

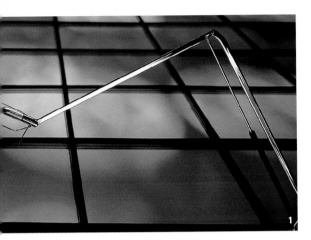

1 "Lifto" by Benjamin Thut for Belux (1984)
Articulated arm light in glass and iron; halogen, 50w; 12v. Transformer.

2 "YaYaHo" by Ingo Maurer for Ingo Maurer (1984)
Track/mirror system; low-voltage halogen, variable wattage. Transformers. *Installation at the Louisiana Museum of Modern Art, Humlebaek, Denmark.*

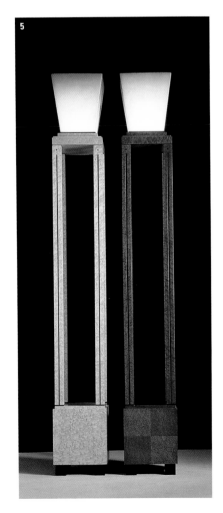

3 "Nordica" by Equipo Santa & Cole for Santa & Cole (1987)
Floor light in metal and wood; tungsten, 100w; standard voltage. Dimmer; adjustable height and shade.

4 "Berenice" by Paolo Rizzatto & Alberto Meda for Luceplan (1985)
Table light in aluminium and pressed glass; halogen, 35w; 12v. Transformer and reflector.

5 "Ingrid" by Michael Graves for Sawaya & Moroni (1987)
Floor light in solid wood with inlaid bird's eye maple veneer or stained mahogany; black lacquer feet and onyx shade; incandescent or halogen, 100w; standard voltage. Dimmer. *Graves' post-modern lights echo classical pillars, with the light as the capital.*

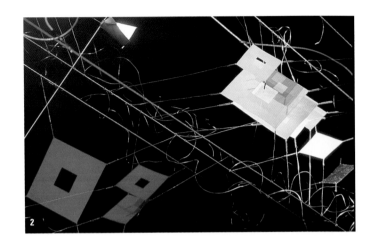

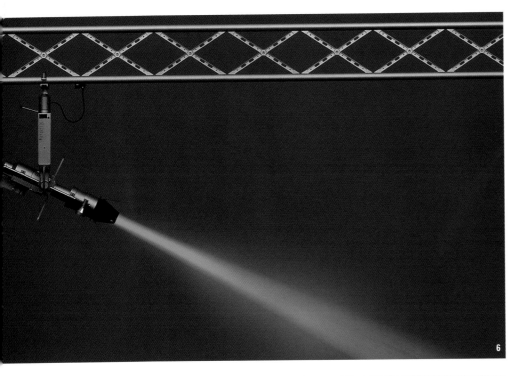

6

7

8

9

10

6 "Emanon High-Performance Spotlight" by Roy Fleetwood for Erco (1980s)
High-performance projector spotlights in metal and glass; various light sources, wattages, and voltages; track fixing.

7 "Costanza" by Paolo Rizzatto for Luceplan (1986)
Table light in aluminium, silk-screened shade; tungsten, 150w; standard voltage. Telescopic, sensorial dimmer.

8 "Tolomeo" by Michele de Lucchi & G Fassima for Artemide (1987)
Table light in aluminium; incandescent (100w), halogen (50w), or fluorescent (13w); standard voltage.

9 "On Taro" and "On Giro" by Keith Haring for Kreon (1987)
Table lights with etched glass panels held by rock bases; electro-luminescent panels (phosphoric effect derived from crystals); 1w; standard voltage.

10 "Butterfly" by Afra and Tobia Scarpa for Flos (1985)
Floor light with etched glass and metal with fire-retardant fabric screen; halogen, 300w; standard voltage. Diffuser and dimmer.

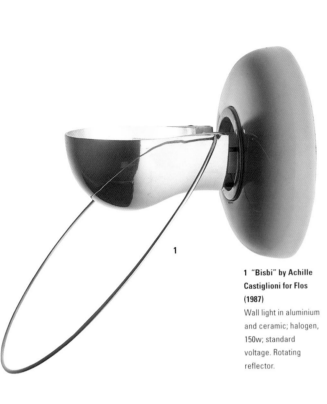

1 "Bisbi" by Achille Castiglioni for Flos (1987)
Wall light in aluminium and ceramic; halogen, 150w; standard voltage. Rotating reflector.

2 "Taraxacum" by Achille Castiglioni for Flos (1988)
Ceiling light in aluminium and glass; incandescent, 60 x 25/40w; standard voltage.

3 "One from the Heart" by Ingo Maurer for Ingo Maurer (1989)
Table light in red and black metal and plastic with adjustable glass mirror; dichroic, 50w; 12v. "Touchtronic" electronic transformer-sensor-dimmer.

4 "Sillaba" by Achille Castiglioni for Fontana Arte (1989)
Ceiling light in thermoplastic technopolymer, glass diffuser; halogen, 75w; 12v.

5 "Trybeca" by Bernhard Dessecker for Ingo Maurer (1989)
Wall light in plastic and aluminium, with Plexiglas; halogen, 2 x 75w; standard voltage. Reflectors. *Pleated metal that looks like corrugated cardboard shields the standard fitting. Trybeca, which plugs into any wall outlet, is completely adjustable; it moves up and down and swivels on its fixture.*

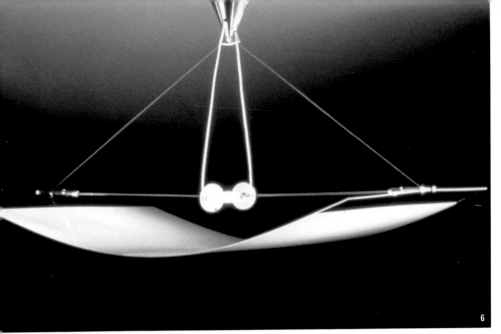

6 "Velo 2791" by
Franco Raggi for
Fontana Arte (1988)
Ceiling light in sheet
glass with galvanized
steel supports;
halogen, 300w;
standard voltage.

7 "Tango" by Stephan
Copeland for Flos
(1989)
Table light in
technopolymer;
halogen, 50w; 12v.
Transformer.

8 "Cornice" by Ramon
Bigas & Pep Sant for
Luxo (1989)
Modular wall system
in aluminium and
coloured methacrylate;
compact fluorescent/
metal-halide, various
wattages; standard
voltage. Ornamental
profile/track fixing,
flexible joint; diffuser.
*An innovative
aluminium track that
runs along walls like
an old-fashioned
moulding and lights
bulbs (lamps)
wherever they are
clipped in.*

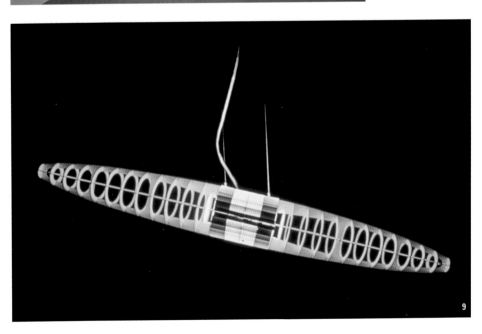

9 "Titania" by Alberto
Meda & Paolo Rizzatto
for Luceplan (1989)
Ceiling light in silver
aluminium with
removable coloured
polycarbonate filters;
halogen, 250w;
standard voltage.
Counterweighted for
vertical adjustment.

10 "Arà" by Philippe
Starck for Flos (1988)
Table light in
chromium-plated
metal; halogen, 35w;
12v. Transformer.

1

2

3

1 "Miss Sissi" by Philippe Starck for Flos (1991)
Table light in technopolymer plastic; incandescent, 60w; standard voltage. Diffuser.

2 "Nyhavn" by Alfred Homann for Louis Poulsen (1993)
Ceiling light in sand-blasted or white-painted cast silumin and polycarbonate; incandescent or fluorescent, 150w; standard voltage. Diffuser.

3 "P1250" by Leonardo Marelli for Estiluz (1991)
Floor light in metal and transparent glass; halogen, 200w; standard voltage. Integrated electronic dimmers.

1990s classics

Conventional-looking light fittings with a shade atop a stemmed base hide the fact that lights became technologically more interesting in the 1990s. In-built controls like touch-sensitive switches freed lights from wires and external controls. Designers also explored light-emitting materials to shape the fitting. For example, Philippe Starck's "Miss Sissi" is a familiar bedside companion, yet the way that it evenly diffuses light through both the shade and the base outstrips other performers. These light fittings may look as though they were designed at the turn of the 20th century rather than at the start of the 21st, but the minute that they are switched on their performance differs dramatically from earlier designs. New ways of transmitting light are also important to today's designers, from Ross Lovegrove's solar-powered garden lights to Jean Nouvel's transformation of 3M's plastic film, originally designed for highway signs, into a floor light.

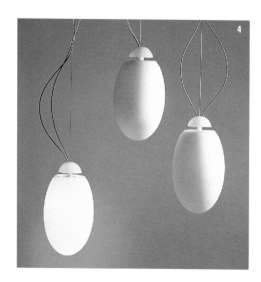

4

4 "Brera" by Achille Castiglioni for Flos (1992)
Ceiling light in glass and chromium-plated or painted metal; incandescent, 100w, or fluorescent, 23w; standard voltage. Three different heights.

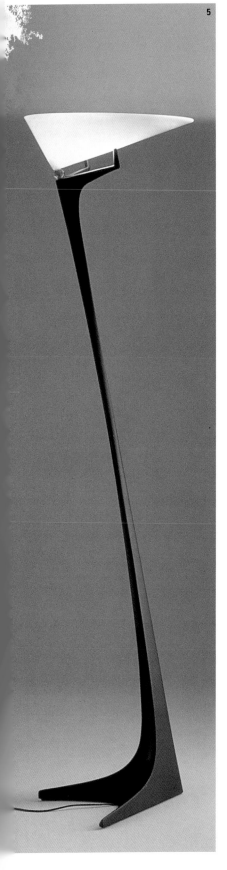

5 "Montjuic" by Santiago Calatrava for Artemide (1990)
Floor light in black-painted foamed polyurethane and white opaline methacrylate; halogen, 400w; standard voltage. Dimmer.

6 "Ursa Major" by Vico Magistretti for Nemo (1993)
Ceiling light in aluminium and moulded glass; halogen or fluorescent, 100w; 20v.

7 "Rosy Angelis" by Philippe Starck for Flos (1991)
Floor light in aluminium and fabric; incandescent, 150w; standard voltage. Electronic twist dimmer.

8 "Lucellino" by Ingo Maurer for Ingo Maurer (1992)
Wall light in glass, brass, plastic, and feather; special incandescent, 50w; 24v. Touch-tronic dimmer/ transformer.

9 "Open light" by Kaori Shimanaka for Koisumi International Lighting Design (1991)
Table light.

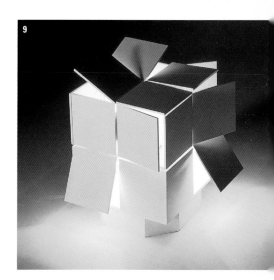

1 "Fuscia" by Achille Castiglioni for Flos (1996)
Pendant light in blown glass, silicone, and plastic; incandescent, 12 x 40w; standard voltage. Diffuser cones.

2 "Zette" by Ingo Maurer for Ingo Maurer (1997)
Ceiling light in stainless steel, satin-frosted glass, and Japanese paper; halogen, 250w; 230/125v (top); and halogen PAR 30, 75w/230v (bottom).

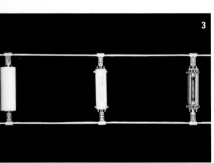

3 "AT2FTR" by Piero Castiglioni for Fontana Arte (1997)
Ceiling/wall light in steel and chrome and clear or frosted Pyrex glass; linear halogen, 300w (up to maximum of 2500w); standard voltage. Steel track fixing.

4 "Romeo Moon" by Philippe Starck for Flos (1996)
Pendant light in pressed glass with sanded inner glass diffuser and steel wires; incandescent, halogen, 150w; standard voltage. *The most famous 1990s designer, Starck uses organic forms and modern materials to reinvent traditional products such as this pendant shade.*

5 "Tricycle" by Shui Kay Kan for SKK (1994)
Table light in zinc-plated steel, plastic, and aluminium; halogen, 40w; standard voltage. Customized images available.

7 "Stresa G Sospensione" by Asahara Sigheaki for Lucitalia (1996)
Suspension light in die-cast aluminium; halogen, 150w; standard voltage. Glass diffuser.

8 "Kabokov" by Ingo Maurer for Ingo Maurer (1993)
Ceiling light with metal canopy, polished glass reflector, and lead counterweight; halogen, 250w; standard voltage.

6 "Orchestra" by Alberto Meda and Paolo Rizzatto for Luceplan (1995)
Wall light in aluminized steel; compact fluorescent, 10w, or halogen, 60w; standard voltage. Reflector, diffuser.

9 "Beret" by Marianne Tuxen for Louis Poulsen (1995)
Table light in white polycarbonate and stainless steel; compact fluorescent, 2 x 18w; standard voltage.

10 "Jack Light" by Tom Dixon for Eurolounge (1996)
Multifunction (floor/table, stackable) light in polyethylene; incandescent, 40w; standard voltage. Also luminescent version. *Made from recycled plastics, the same sort used for traffic bollards, "Jack Light" is tough enough to double as a stool and stacks as well.*

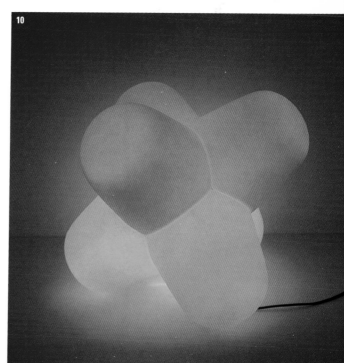

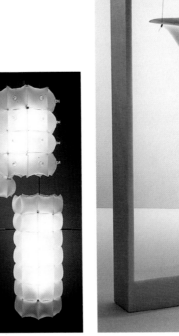

3 "(Y1) Passagi" by
Andrea Branzi for
Design Gallery Milano
(1998)
Free standing light in
porcelain.

1 "Assia" by
Francesco Rota for
O Luce (1998)
Ceiling or floor light in
black-lacquered metal
with Plexiglas diffuser;
incandescent, 4 x 60w;
standard voltage.

2 "Grid" by Katrien
Van Liefferinge for
Katrien Van
Liefferinge (1997)
Inflatable ceiling light
in natural PVC and
polished aluminium;
incandescent, 2 x 60w;
standard voltage.

4 "Oskar" by
Ingo Maurer for
Ingo Maurer (1998)
Table light/bookrest in
anodized aluminium;
halogen, 50w; 12v.
Electronic safety
transformer (125/230v).
Flexible metal arm to
adjust reflector.

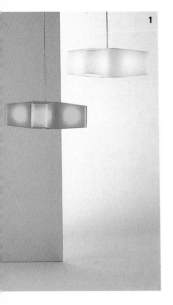

US showroom
IE Architectural
233 Cleveland Avenue
Highland Park, NJ 08904
USA
tel: +1 732 745 5858
fax: +1 732 745 9710

Luxo Italiana SpA
Via delle More, 1
24030 Presezzo (BG)
Italy
tel: +39 35 464818
fax: +39 35 464817

Matsushita Electric Industrial Co Ltd
1006, Oaza Kadoma
Kadoma City
Osaka
Japan
tel: +81 6 908 1121
fax: +81 6 908 2351
web: www.mei.co.jp

US office
Matsushita Electric Corporation of America
1 Panasonic Way
Secaucus, NJ 07094-2917
USA
tel: +1 201 348 7000
fax: +1 201 348 8378
web: www.panasonic.com

Mazzega A V srl
Via Vivarini, 3
30141 Murano (VE)
Italy
tel: +39 41 736677
fax: +39 41 739939

Memphis/Post Design
Via Olivetti, 9
20010 Pregnana Milanese (MI)
Italy
tel: +39 29 329 0663
fax: +39 29 359 1202
e-mail: memphis@memphis-milano.com
web: www.memphis-milano.com

Metalarte SA
Avda de Barcelona, 4
08970 Sant Joan Despi
Barcelona
Spain
tel: +34 3 477 0069
fax: +34 3 477 0086

UK distributor
Mr Light
275 Fulham Road
London SW10 9PZ
England
tel: +44 171 351 1487
fax: +44 171 351 3484

Nemo srl
Via Busnelli, 1
20036 Meda (MI)
Italy
tel: +39 36 234 0570
fax: +39 36 234 0599
e-mail: nemolite@tin.it
web: web.tin.it/nemolite

UK distributor
Atrium Ltd
22–24 St Giles High Street
London WC2H 8LN
England
tel: +44 171 379 7288
fax: +44 171 240 2080

Nimbus GmbH
Rosenbergstrasse, 113
70193 Stuttgart
Germany
tel: +49 711 630055
fax: +49 711 630066

UK distributor
GFC Lighting Ltd
Westminster Business Square
London SE11 5JH
England
tel +44 171 735 0677
fax: +44 171 793 0122

US distributor
Alexandra Hafner
9 Lloyd Cours
Nappa, CA 94558
USA
tel: +1 707 226 6387
fax: +1 707 226 6829

Noguchi/Akari Associates
32-37 Vernon Boulevard
Long Island City, NY 11106
USA
tel: +1 718 721 2308
fax: +1 718 278 2348
e-mail: akari@noguchi.org
web: www.noguchi.org

O Luce
Via Cavour, 52
20098 San Giuliano (MI)
Italy
tel: +39 29 849 1435
fax: +39 29 849 0779
e-mail: oluce@galactica.it

UK distributor
Interior Marketing
2 Woods Cottages
Hatfield Broad Oak
Bishop's Stortford
Herts CM22 7BU
England
tel: +44 1279 718151
fax: +44 1279 718152

Pallucco Italia SpA
Via Azzi, 36
31040 Castagnole di Paese
Italy
tel: +39 42 243 8800
fax: +39 42 243 8555

UK distributor
Viaduct
1–10 Summer Street
London EC1R 5BD
England
tel: +44 171 278 8456
fax: +44 171 278 2844

Pandul
Vaerkstedsvej, 17
2500 Valby
Denmark
tel: +45 36458303
fax: +45 36458203

UK distributor
Schiang UK
355 Portobello Road
London W10 5SA
England
tel: +44 181 960 9028
fax: +44 181 968 0828
e-mail: sales@schiang.com
web: www.danish-design.com

Post Design see **Memphis/Post Design**

Santa & Cole
Santisima Trinidad del Monte, 10
08017 Barcelona
Spain
tel: +34 3 417 8455
fax: +34 3 418 7058

UK distributor
John Arend
69 Shirland Road
Twickenham
Middx TW1 4HB
England
tel: +44 181 892 2278
fax: +44 181 744 1375

US distributor
Resolute
1013 Stewart Street
Seattle, WA 98101
USA
tel: +1 206 343 9323
fax: +1 206 343 9322

Sawaya & Moroni SpA
Via Andegari, 18
20121 Milan
Italy
tel: +39 2 863951
fax: +39 2 86464831

SKK Lighting
34 Lexington Street
London W1R 3HR
England
tel: +44 171 434 4095
fax: +44 171 287 0168
e-mail:skk@easynet.co.uk
web: www.easynet.co.uk.skk/

Daniel Swarovski & Co
Swarovskistrasse, 30
Postfach 15
A-6112 Wattens
Austria
tel: +43 5224 5000
fax: +43 5224 52335

UK office
Swarovski UK Ltd
Perrywood Business Park
Salfords
Surrey RH1 5JQ
England
tel: +44 1737 856800
fax: +44 1737 856880
web: www.swarovski.com

US office
Swarovski Consumer Goods Ltd
1 Kenney Drive
Cranston, RI 02920-4400
USA
tel: +1 401 463 6400
fax: +1 401 463 5257

Technolumen GmbH & Co KG
Lötzener Strasse, 2–4
D-28207 Bremen
Germany
tel: +49 421 444016
fax: +49 421 4986685

Venini SpA
Fondamenta Vetrai, 50
30141 Murano (VE)
Italy
tel: +39 41 739955
fax: +39 41 739369
e-mail: venini@venini.it
web: www.venini.it

Woka Lamps Vienna
Palais Breuner Singerstrasse, 16
A-1010 Vienna 1
Austria
tel: +43 1 513 2912
fax: +43 1 513 8505

Ycami Edizioni
Via Provinciale, 31
22060 Novedrate (CO)
Italy
tel: +39 31 351 0111
fax: +39 31 351 0400

BULB (LAMP) MANUFACTURERS

B-Ticino SpA
Via Messina, 38
20154 Milan
Italy
tel: +39 2 34801
fax: +39 2 3480707
web: www.bticino.it

UK distributor
Lagrad Electric
Foster Avenue
Woodside Park
Dunstable
Beds LU5 5TA
England
tel: +44 1582 676767
fax: +44 1582 676771

Concord Lighting Ltd
174 High Holborn
London WC1V 7AA
England
tel: +44 171 497 1400
fax: +44 171 497 1404

US distributor
Sylvania Lighting International
6600 N Andrews Avenue
Suite 240
Fort Lauderdale, FL 33309
USA
tel: +1 954 776 1606
fax: +1 954 491 1338

Helvar Ltd
Hawley Mill
Hawley Road
Dartford
Kent DA2 7SY
England
tel: +44 1322 282258
fax: +44 1233 282259

Illumna Lighting Ltd
24–32 Riverside Way
Uxbridge
Middx UB8 2YF
England
tel: +44 1895 272275
fax: +44 1895 270024

Irideon (Lighting Technology Group)
2 Tudor Estate
Abbey Road
Park Royal
London NW10
England
tel: +44 181 965 6800
fax: +44 181 965 0970

US office
Irideon Inc
3030 Laura Lane
Middleton, WI 53562
USA
tel: +1 608 831 4116
fax: +1 608 836 1736

Lumiram
PO Box 297
Mamaroneck, NY 10543
USA
tel: +1 914 698 1205
fax: +1 914 698 5468

Lutron EA Ltd
Lutron House
6 Sovereign Close
London E1 9HW
England
tel: +44 171 702 0657
fax: +44 171 480 6899

Maris Ltd
Little Beachfield Farm
Heron Hill Lane
Meopham
Kent DA13 3QL
England
tel: +44 1474 812999
fax: +44 1474 812777

Meyer Lighting Co
PO Box 691796
Orlando, FL 32869-1796
USA
tel: +1 407 351 5474
fax: +1 407 363 4886

UK distributor
Commercial Lighting Systems Ltd
Units 16–17
Park Gate Business Centre
Southampton
Hants SO31 1FQ
England
tel: +44 1489 581002
fax: +44 1489 576262

Osram Ltd
PO Box 17
East Lane
North Wembley
Middx HA9 76PG
England
tel: +44 181 904 4321
fax: +44 181 901 1222

Philips Electronics BV
PO Box 218
5600 MD Eindhoven
The Netherlands
tel: +31 40 279 1111
fax: +31 20 597 7150
web: www.philips.com

UK office
Philips Lighting Ltd
The Philips Centre
420–430 London Road
Croydon CR9 3QR
England
tel: +44 181 665 6655
fax: +44 181 665 5100

US office
Philips Electronics North America
1251 Avenue of the Americas
New York, NY 10020
USA
tel: +1 212 536 0500
fax: +1 212 536 0559

Reggiani SpA Illuminazione
Via della Misericordia, 33
20057 Vedano al Lambro (MI)
Italy
tel: +39 49 1021
fax: +39 49 2714

UK office
Reggiani Ltd Lighting
12 Chester Road
Borehamwood
Herts WD6 1LT
England
tel: +44 181 95030855
fax: +44 181 2073923

US office
Reggiani USA Inc Illumination
108 South Water Street
Newburgh, NY 12550
USA
tel: +1 914 565 8500
fax: +1 914 561 1130

Staff Lighting
Unit 4 – The Argent Centre
Pump Lane
Hayes
Middx UB3 3BL
England
tel: +44 181 569 3639
fax: +44 181 573 3560

Thorn Lighting
3 King George Close
Eastern Avenue West
Romford
Essex RM7 7PP
England
tel: +44 1708 766003
fax: +44 1708 776269

3M
3M Center
St Paul, MN 55144-1000
USA
tel: +1 651 733 1110
fax: +1 651 733 9973

UK office
3M United Kingdom plc
PO Box 1
Market Place
Bracknell
Berks RG12 1JU
England
tel: +44 1990 360036
fax: +44 1344 858278

Zumtobel
Schweizerstrasse, 30
A-6850 Dornbirn
Austria
tel: + 43 1258 26010
fax: + 43 5572 20721

UK office
Zumtobel Lighting Systems Ltd
Unit 5 – The Argent Centre
Pump Lane
Hayes
Middx UB3 3BL
England
tel: +44 181 573 3556
fax: +44 181 573 3560

US office
Zumtobel Lighting Inc
141 Lanza Avenue
Building 16D
Garfield, NJ 07026
USA
tel: +1 914 691 6262
fax: +1 914 691 6289

FOUNDATIONS AND FEDERATIONS

Domus Academy
Via Savona, 97
20144 Milan
Italy
tel: +39 2 47719455
fax: +39 2 422525

The Getty Center
1200 Getty Center Drive
Suite 400
Los Angeles, CA 90049-1681
USA
tel: +1 310 440 7360
fax: +1 310 440 7722

Guggenheim Museum
1071 Fifth Avenue
New York, NY 10128
USA
tel: +1 212 423 3500
fax: +1 212 423 3650

International Dark-Sky Association
3225 North First Avenue
Tucson, AZ 85719
USA
tel: +1 520 293 3198
fax: +1 520 293 3192
web: www.darksky.org

Isamu Noguchi Foundation
The Isamu Noguchi Garden Museum
32-37 Vernon Boulevard
Long Island City, NY 11106
USA
tel: +1 718 721 1932
fax: +1 718 278 2348

Lighting Industry Federation Ltd
207 Balham High Road
London SW17 7BQ
England
tel: +44 181 675 5432
fax: +44 181 673 5880

Ruskin Library
University of Lancaster
Lancs LA1 4YW
England
tel: +44 1524 593587
fax: +44 1524 593580

Sir John Soane's Museum
13 Lincoln's Inn Fields
London WC2A 3BP
England
tel: +44 171 405 2107
fax: +44 171 831 3597

ARCHITECTS, DESIGNERS AND ARTISTS

Tadao Ando Architect & Associates
5–23 Toyosaki 2-Chome
Osaka
Japan 531-0072
tel: +81 6 375 1148
fax: +81 6 374 6240

Ron Arad
62 Chalk Farm Road
London NW1
England
tel: +44 171 284 4963
fax: +44 171 379 0499

BDP Lighting
16 Gresse Street
London W1A 4WD
England
tel: +44 171 631 4733
fax: +44 171 631 0393

Mario Botta
Via Chiani, 16
6904 Lugano
Switzerland
tel: +41 91 972 8625
fax: +41 91 970 1454

Andrea Branzi
Via Moscova, 43
20121 Milan
Italy
tel: +39 2 659 2227
fax: +39 2 657 1676

Arnold Chan
Isometrix Lighting & Design
8 Glasshouse Yard
London EC1A 4JN
England
tel: +44 171 253 2888
fax: +44 171 253 2899

Sophie Chandler
Unit 13a (3rd floor)
246 Stockwell Road
London SW9
England
tel & fax: +44 171 582 2676

Diana Edmunds
52 Park Hall Road
London SE1 8BW
England
tel: +44 171 608 7900
fax: +44 171 608 7901

Mark Fisher Studio
51 Wharton Street
London WC1X 9PA
England
tel: +44 171 713 1884
fax: +44 171 713 1885

Sir Norman Foster
Foster & Partners
22 Hester Road
London SW11 4AN
England
tel: +44 171 738 0455
fax: +44 171 738 1104

Future Systems
21c Conduit Place
London W2 1HS
England
tel: +44 171 723 4141
fax: +44 171 723 1131

Frank Gehry
Frank O Gehry Associates
1520B Cloverfield Blvd
Santa Monica, CA 90404
USA
tel: +1 310 828 6088
fax: +1 310 828 2098

Piers Gough
CZWG
17 Bowling Green Lane
London EC1R 0BD
England
tel: +44 171 253 2523
fax: +44 171 250 0594

Michael Graves
341 Nassau Street
Princeton, NJ 08540
USA
tel: +1 609 924 6409
fax: +1 609 924 1795

Mark Guard Architects
161 Whitfield Street
London W1P 5RY
England
tel: +44 171 380 1199
fax: +44 171 387 5441
e-mail: mga@markguard.co.uk
web: www.markguard.co.uk

Eva Jiricna Architects
Sun Court House
18–26 Essex Road
London N1 8LN
England
tel: +44 171 704 7300
fax: +44 171 226 8903

King & Miranda Associati
Via Forcella, 3
20100 Milan
Italy
tel: +39 2 839 4963
fax: +39 2 836 0735

Katrien Van Liefferinge
13 Armstrong Street
Leeds
Yorks LS28 5BZ
England
tel & fax: +44 113 2570241

Light & Design Associates
Gladstone Studios
Rock Hill
Chelsfield
Kent BR6 7PJ
England
tel: +44 1959 534425
fax: +44 1959 533641

Jeremy Lord
see **The Colour Light Co Ltd**
(see Light Fitting Manufacturers)

Ross Lovegrove
Studio X
21 Powis Terrace
London W11 1JN
England
tel: +44 171 229 7104
fax: +44 171 229 7032

Richard MacCormac
MacCormac, Jamieson & Pritchard
9 Heneage Street
London E1 5LJ
England
tel: +44 171 377 9262
fax: +44 171 247 7854

Peter Marino & Associates
150 East 58 Street
New York, NY 10022
USA
tel: +1 212 752 5444
fax: +1 212 759 3727

Javier Mariscal
Estudio Mariscal
Pellaires, 30–38
8819 Barcelona
Spain
tel: +34 3 303 3420
fax: +34 3 266 2244

Rick Mather
123 Camden High Street
London NW1 7JR
England
tel: +44 171 284 1727
fax: +44 171 267 7826

Richard Meier & Partners
475 10th Avenue
New York, NY 10018-1120
USA
tel: +1 212 967 6060
fax: +1 212 967 3207

Jean Nouvel
(architecture)
10, cité Angouleme
F-75011 Paris
France
tel: +33 1 49238383
fax: +33 1 43148110

(design)
41, rue Francs Bourgeois
F-75004 Paris
France
tel: +33 1 44610110
fax: +33 1 44610120

John Pawson
70–78 York Way
London N1 9AG
England
tel: +44 171 837 2929
fax: +44 171 837 4949

I M Pei
Pei Cobb Freed & Partners
600 Madison Avenue
New York, NY 10022
USA
tel: +1 212 751 3122
fax: +1 212 872 5443

Julian Powell-Tuck
Powell Tuck Associates Ltd
14 Barley Mow Passage
London W4 4PH
England
tel: +44 181 747 9988
fax: +44 181 747 8838

Nico Rensch
Cheynes Farm
Cottered
Herts SG9 9QE
England
tel: +44 1763 281566
fax: +44 1763 281577

Richard Rogers Architects Partnership
Thames Wharf
London W6 9HA
England
tel: +44 171 385 1235
fax: +44 171 385 8409

Bruno & Sandra Rohrbach
Langhagstrasse, 2
CH-4410 Liestal
Switzerland
tel: +41 61 921 3221
fax: +41 62 921 3125

Agence Philippe Starck
27, rue Pierre Poli
92130 Issy-les-Moulineaux
France
tel: +33 1 41088282
fax: +33 1 41089665

Claudio Silvestrin Limited
392 St John Street
London EC1V 4NN
England
tel: +44 171 323 6564
fax: +44 171 833 1861

Jonathan Speirs & Associates
Well Court Hall
Dean Village
Edinburgh EH4 3BE
Scotland
tel: +44 131 226 4474
fax: +44 131 220 5331

Speirs & Major
14 West Central Street
London WC1A 1JH
England
tel: +44 171 240 4042
fax: +44 171 240 4046

Sally Storey
Lighting Design International
Zero Ellaline Road
London W6 9NZ
England
tel: +44 171 381 8999
fax: +44 171 385 0042

Patrick Woodroffe
Radford Farm
Northend
Bath
Avon BA1 8ES
England
tel: +44 1225 852646
fax: +44 1225 852009

OTHER USEFUL ADDRESSES

Electrical Contractors Association
34 Palace Court
Bayswater
London W2 4HY
England
tel: +44 171 313 4800
fax: +44 171 221 7344

National Electrical Contractors Association
3 Bethesda Metro Center
Suite 1100
Bethesda, MD 20814
USA
tel: +1 301 657 3110
fax: +1 301 215 4500
web: www.necanet.org

Index

Nonie Niesewand would like to thank everybody who helped throw light on the complex and fascinating subject of lighting, and in particular:

Ingo Maurer, who turned me onto the subject

The Domus Academy in Milan, whose course on the subject was both poetic and pragmatic

All the architects and lighting consultants featured in this book, every one of whom shared their specialist knowledge. They are: Sally Storey of Lighting Design International; Richard MacCormac; Philippe Starck; Jonathan Speirs and Mark Major; Future Systems; Mark Guard; Michael Graves; Claudio Silvestrin; Eva Jiricna; Rick Mather; Tadao Ando; Julian Powell-Tuck; John Pawson; Ross Lovegrove and Misha Miller; Peter Marino; Nico Rensch.

The lighting manufacturers and their agents, and in particular: John Cook of Flos Lighting in the UK; John Roake of Catalytico; Erco Lighting; the Design Lighting Federation.

AKG, London 12 bottom right /Collection Hahnloser, Bern 21 top right; /©DACS 1998 12 top; /Gebrauchsgraphik, Berlin, July 1929 12 bottom left; /Mies van der Rohe/Erich Lessing 19 left; /Museo del Prado, Madrid 11 bottom; /Vienna, Galerie Wuerthle/Erich Lessing 29 bottom **Alivar** 143 (4) **Andrea Branzi** 16 **Anglepoise Ltd** 143 (6) **Arcaid** 30; /Early Carter/Belle/Designer: Philippe Starck, Delano Hotel, Miami, Florida, USA 126, 127; /John Edward Linden/Architect: Richard Meier 93; /Mark Fiennes/Architect: Michael Graves 132; /Nicholas Kane/Architect: Nico Rensch 91 top right; /Richard Bryant/Architect: Eva Jiricna 94; /Richard Bryant/Architect: John Pawson Architects 53 right; /Richard Bryant/Architects: MacCormac, Jamieson and Prichard 68; /Richard Bryant/Architect: Richard Rogers 27 top right; /Richard Bryant/Architect: Tadao Ando 108 top and bottom, 190; /Richard Bryant/Architect: Eva Jiricna

Co/Jeremy Lord 22 bottom right **Concord** 50 centre left, 92 right, lower centre **Concord Sylvania**/John Edward Linden/Architect: Powell Tuck & Associates 88 **The Conde Nast Publications Ltd**/Tim Clinch/Designer: David Hicks 66 left, 67 **Corbis UK Ltd**/Ansel Adams Publishing Rights Trust 6; /Kit Kittle 51 right; /Macduff Everton 32 bottom left **Design Gallery Milano** 162 (3), 163 (7) **Diana Edmunds** 57 **DMD/Droog**/ Erik Jan Kwakkel/Design: Arnout Visser 61 left, /Marsel Loermans/Design: Dick van Hoff 163 (6) **Donghia** 137 (9) **Dumoffice** 54 centre, 163 (9) **Edison Price** 92 left **Erco** 11 top; /31 top, 90 (b), 92 (c), 153 (6), 155 (6) **Estiluz** 158 (3) **Flos** 41, 44 bottom right, 45 left, 55 bottom right, 58 bottom right, 59 centre, 137 (7, 8), 144 (2), 145 (7, 8, 9), 146 (1), 148 (3), 149 (6, 9), 150 (2, 4), 151 (7, 9), 152 (2), 153 (7), 155 (10), 156 (1, 2), 157 (7, 10), 158 (1, 4), 159 (7), 160 (1, 4), back cover, centre **Fontana**

MacCormac, Jamieson and Prichard/ Graham Cook 91 top left; /Hollands Licht/Rogier van der Heide 92 bottom right; /Peter Durrant 69 top, 69 bottom **Mainstream**/Ray Main 38–39, 62–63 **Mark Guard & Associates**/Alan Williams 98–99 **Mazzega** 139 (6) **Metalarte** 148 (5) **Memphis/Post Design**/Studio Azzurro 54 bottom left, 152 (1, 3), 153 (8), 163 (5) **Mitchell Beazley**/Neil Mersh 161 (10) /Tim Clinch/ Architect: Michael Graves 100, 101, 133; /Tim Clinch/Designer: Philippe Starck 80, 81 left, 81, 82 right, 83 left and right **The Moviestore Collection**/British Lion 17 top; /Columbia 36 top right **Nemo** 159 (6) **Nimbus** 46 left, 54 bottom right **O Luce** 139 (9), 148 (1), 151 (8), 162 (1), 163 (8) **Palluco Italia** 141 (5) **Pandul** 147 (4), 149 (7), 153 (4) **Paul Rocheleau**/© ARS, NY and DACS, London 1998 back cover, left and right, 9 **Philips** 56 bottom right **Powell-Tuck & Associates**/ Henry

Acknowledgements

(originally Ove Arup) 130, 131; /Richard Bryant/Architects: Gale & Prior 28 bottom; /Richard Bryant/Sir John Soane 64 left and right, 65; /Richard Glover 22 bottom left **Architectural Association Picture Library**/Alan Chandler/Foundation Le Corbusier/ADAGP, Paris & DACS, London 1998 31 left **Artek** 140 (3) **Artemide** 1, 2, 17 bottom, 58 top right, 40, 44 top left, 48 centre and bottom left, 50 top left, 60 right, 138 (1), 144 (4), 146 (2), 147 (6, 7), 148 (2), 149 (8), /Aldo Ballo 150 (3), 155 (8), 159 (5) **Athena** 56 top left, centre left **Axiom Photographic Agency**/James Morris front cover, 18 top; /James Morris/Architect: John Pawson 19 right, 114 left, 114–15, 124 left, 124–25; /James Morris/Architect: Claudio Silvestrin (Riverside Apartments: Foster & Partners) 4–5, 102–103, 103 bottom right and top right **B.Lux** 49 centre right **Barovier & Toso** 136 (1), 137 (6) **Beth Coyne**/Gavin Cochrane 73 top right, 74 left; /Gavin Cochrane/Designer: Sally Storey 72 top **Belux** 43, 44 (centre), 45 right, 49 top right, 50 bottom left, 53 left, 54 top, 55 top, 134–35, 153 (9), 154 (1), 163 (10) **Bisazza**/Santi Caleca 27 btm left **Blauet** 136 (3) **Bridgeman Art Library**/David Ker Fine Art, London, UK/© Courtesy of the Artist's Estate 24 top right; /Kunstmuseum Bern, Switzerland/© DACS 1998 27 bottom right; /Private Collection 29 top; /Private Collection/Peter Willi 34 bottom; /Saatchi Collection, London, UK/© ARS, NY and DACS, London 1998 32 top; /Skagens Museum, Denmark 21 bottom right **BTicino** 60 bottom left **ClassiCon** 141 (6), 142 (1) **Colour Light**

Arte 143 (8), 144 (1, 5), 147 (5), 156 (4), 157 (6), 160 (3) **Foscarini** 44 bottom left, 48 top right, 137 (5), 138 (3), 139 (8) **Foster & Partners**/ Ian Lambot 21 bottom left; /Richard Davies 91 bottom **Hiroyuki Hirai** 56 bottom left **Iketrade**/Ingo Maurer 112, 113; /Maria Grazia Branco/ Architect: Ross Lovegrove 116, 117 **Images Colour Library Limited** 26 left and right, 32 bottom centre, 34 top, 35 bottom **Ingo Maurer** 13, 48 (top left), 84, 85; /Friedrich Busam 86 and 87; 148 (4), 150 (1), 151 (5, 6), 154 (2), 156 (3, 5), 159 (8), 160 (2), 161 (8), 162 (4) **The Interior Archive**/Schulenburg/Architect: Nico Rensch 128–29 **Isamu Noguchi Foundation Inc**/Kevin Noble 147 (8, 9) **James Wojcik** 42, 46 right, 50 right, 52 **John Cullen Lighting** 73c **Josep Lluscà** 48 bottom right **Jonathan Speirs & Associates** 76, 77, 78, 79 **Kate Wild** 73 left, 74 right, top to bottom **Katrien Van Liefferinge**/David Wolfenden/T. Campbell 55 left; / **Kobal Collection**/Channel 4/Glasgow Film Board 24 bottom right; /ERA International–91 37 left **Koisumi International Lighting Design** 159 (9) **Kreon**/F. Debras 8, 14–15, 155 (9) **Leucos** 136 (2, 4) **Lighting Industry Federation** (National Lighting Design Awards 1997/98) 23, /Chorley Handford 89 top left, 89 bottom, 90 left **Louis Poulsen** 143 (5), 144 (3), 145 (6), 146 (3), 158 (2), 161 (9) **Luceplan** 44 top right, 49 top left, 58 left, /Leo Torri 59 top left and top right, 61 centre and right; 154 (4), 155 (7), 157 (9), 158 (2), 161 (6) **Lucitalia** 153 (5), 161 (7) **Luxo** 49 bottom left, 143 (7), 157 (8)

Bourne 110–11,118–19 **Redferns**/Mark Fisher 31 bottom right; /P. Ford 31 centre right; **Resolute/Manifesto** endpapers; **Rex Features** 32 bottom right; /Christo and Jeanne Claude 28 top **Richard Davies** /Architects: Future Systems 47, 96, 97; /Architects: Powell-Tuck & Associates 90 right; **Robert Harding Picture Library**/Adam Woolfitt/ Alvar Aalto 25 right; /F. Jalain/Explorer 18 bottom; /James Merrell/Homes and Gardens, IPC Magazines Ltd/Designer: Sally Storey 70 left and right, 71 top, 71 bottom, 72 bottom, 73 bottom, 75; /Kim Hart/ Foundation Le Corbusier/ADAGP, Paris & DACS, London 1998 22 top **Ron Arad Associates**/C. Kicherer 25 left **Ronald Grant Archive**/© Lucas Film 11 centre; /Courtesy MCM 36 bottom; /Courtesy The Ladd Company 27 top left; /Courtesy Tomson Films 37 right; /Courtesy WFA 36 top left **Santa & Cole** 154 (3) **Sawaya & Moroni**/Marco Schillaci154 (5) **SKK**/Shiu-Kay Kan 160 (5) **Sophie Chandler**/J Pilkington 138 (2) **Swarovski** 33 **Technolumen** 140 (4), 142 (2) **Tim Street-Porter**/Luis Barragan 10 **Venini** 139 (4, 5, 7) **View**/Architects: Peter Marino & Associates 22 top; /Chris Gascoigne/Architects: Peter Marino & Associates 120–21; /Dennis Gilbert/Architect: Frank Gehry 92 top right; /Dennis Gilbert/Architect: Rick Mather 106–7, 122, 123; /Dennis Gilbert/ Architect: Eva Jiricna 104, 105; /Architects: Peter Marino & Associates 20 left; /Architect: Frank Gehry 35 top **Wagenfeld** 143 (4) **Walter Gardiner** 51 left **Woka** 140 (1, 2), 141 (7, 8, 9), 142 (3) **Ycami** 24 left **Yigal Gawze** 56 top right, 89 right